Praise for *The Wild Path Home*

Without question, this is the finest book on raising children I have ever seen. Today's child emerges into a world of strife, loss, and fear. Regenerating life depends on transforming our beliefs and dictums about how to raise a child. Nature will determine if we are worthy to be here on Earth. Every child deserves to encounter, know, and love nature in all its intelligence, complexity, and magnificence. Such a child will realize they are nature and can transform the future for all children.

— Paul Hawken, author, *Carbon*, *Regeneration*, and *The Book of Life*

In my consulting around North America, I often refer to the Pathways to Stewardship project as the most comprehensive community-based approach towards raising children to care about and take care of the earth. It's wonderful to see that they have now translated their fine work into this beautiful book—*The Wild Path Home*. I've always been most impressed by their creation of a set of developmental guidelines that provide guidance for parents and educators of all stripes. Presented here, the Landmarks are articulated in a remarkably accessible form and with lavish photography. These are enhanced with quotes for parents and educators about how they've implemented the Landmarks in their work. *The Wild Path Home* provides guidance that is inspired, sane, gentle, and, most important, effective. Grandparents, parents, and educators can use this book to help figure out what to do with children and students today to help preserve the earth over the long haul.

— David Sobel, environmental educator, Professor Emeritus,
Education Department, Antioch University

We need this book more than ever. *The Wild Path Home* provides a joyful abundance of creative ideas for families, schools, youth programs, and whole communities to help children and youth engage with nature in ways that help them become healthy, resilient protectors of the Earth. Grounded in an evidence-based understanding of child-development stages and nature's benefits, offered through a climate impact lens, and informed by Indigenous wisdom, this is my new go-to guide. What a treasure!

— Cathy Jordan, Ph.D., Director of Research, Children & Nature Network

This beautifully illustrated and well-researched book is an invaluable resource for educators, program planners, parents, grandparents, and anyone who wants to help children and youth develop a lifelong connection with the natural world. Readers will be excited about the hundreds of easy-to-implement yet impactful ideas for different settings and seasons. This essential guidebook is truly impressive!

— Kristi Lekies, Ph.D., Associate Professor, Ohio State University

The Wild Path Home offers a timely and compelling guide for parents, educators, and community leaders seeking to nurture resilient, empathetic, and environmentally conscious children. The authors of this book provide a clear pathway for youth in cultivating stewardship and affinity with the natural and built world around us. It emphasizes the vital importance of unstructured outdoor play, offering children the chance to explore, take risks, and connect deeply with nature, all of which contribute to their physical, mental, and emotional well-being. This book equips readers with practical strategies to inspire the next generation to become active and engaged citizens who will foster meaningful relationships with all life. With an emphasis on respect, reciprocity, and responsibility, this book presents a vision of hope for the future.

— Charles Hopkins, UNESCO Chair in Reorienting Education towards Sustainability, York University

The Wild Path Home is a practical, tender, and inspiring resource for anyone who wants to deepen nature connection for children in their care. As a new parent working in the environmental sector, I appreciate how overwhelming it is to be raising a child during an intensifying climate crisis. With clear and joyful activities grounded in research and traditional knowledge, this book is a balm that gets us outside and into relationship with the world around us.

— Brianna Salmon, Executive Director, Green Communities Canada

The Wild Path Home is an invaluable resource for parents and teachers, and youth programs of all kinds. With accessible, engaging activities, it provides adults with a framework and road map to raising nature's stewards of tomorrow—something more vital now than ever before.

— Drew Monkman, nature educator and mentor

This is a wonderful guide for any adults—teachers, parents, grandparents, youth leaders—looking to play their part in mentoring young people who will engage with, celebrate, care for, and help heal our planet. The Landmarks have relevance wherever this vital work is being done.

— Paul Elliott, Professor Emeritus School of Education
and Professional Learning, Trent University

The Wild Path Home presents a compelling vision for how we can reshape the way children grow up, focusing on respect, interconnectedness, and sustainability. This thought-provoking guide, which draws on Indigenous knowledge, offers a roadmap to nurturing a society that prioritizes health—mental, physical, and environmental—through deeper relationships with nature and one another. It is an essential read for educators, parents, community leaders, and anyone interested in the future of our planet and its inhabitants.

— Alison Elwood, Vice President, Ontario Society of Environmental Educators

Many books offer ideas for activities that foster environmental literacy and environmental stewardship. *The Wild Path Home* does that abundantly, thus serving as a valuable resource. For everyone who wants to work with others in their communities to ensure that all children of all ages have meaningful experiences in nature, this book takes a significant step further by presenting a model for collaboration that can inspire and guide similar initiatives in cities and towns everywhere.

— Louise Chawla, co-author, *Placemaking with Children and Youth*

The Wild Path Home is truly a phenomenal resource for parents and educators alike. The profound power nature has on child development is well documented. This book provides a wealth of activities for connecting children with nature from birth and beyond.

— Kathy Warner RECE, Program Manager, Peterborough Child & Family Centres

The Wild Path Home offers a comprehensive and inspirational framework for nurturing a sense of kinship with the Earth in the hearts and minds of children and youth. The book is beautifully written, the content research-based, and the suggestions applicable to various settings. This guide lives up to its name.

— Ruth Wilson, Ph.D., author and retired educator

In *The Wild Path Home*, visionary environmental educators Jacob Rodenburg and Cathy Dueck have provided an incredible tool for teaching young people. As the climate crisis escalates, fostering understanding of the natural world and problem-solving skills for children of all ages becomes critically important, and this book provides a delightful and grounded approach to both.

— Tegan Moss, Executive Director, GreenUP

Jacob Rodenburg and Cathy Dueck, pioneers in childhood environmental education, present a visionary yet practical approach to nurturing children's connection with nature. Their innovative Pathway framework provides a clear progression of experiences that cultivate wonder, respect, and responsibility toward the natural world. *The Wild Path Home* provides an inspiring framework for families, educators, and communities to raise a generation that understands their place in the world.

— Myke Healy, M.Ed., OCT, Assistant Head of Senior School—Teaching & Learning, Social Sciences, Trinity College School, and Chair, Camp Kawartha

The Wild Path Home is timely, as fewer children today have opportunities to become intimate with the land, water, plants, and animals that surround them. Jacob and Cathy give readers a helpful framework based on age-related landmarks and Indigenous ways of being. These landmarks provide hands-on ideas for enjoying time in nature with children, encouraging a sense of stewardship and kinship. I am looking forward to sharing this book with our early learning community.

— Beckie Evans, Investing in Quality Peterborough, RECE

Aside from its evocative title, *The Wild Path Home* is a finely photographed and extensively field-tested treasure for anyone (parents, other relatives, educators, community leaders, and other mentors) who enjoys taking children and teens outdoors, i.e., home. It is a prescient reminder of what really matters in our lives, and the authors detail an impressive array of doable outdoor activities (from carefully handling earthworms to becoming a citizen scientist) for eight age groups from birth to seventeen years of age. This valuable resource is truly a pathway to stewardship and kinship.

— Grant Linney, career outdoor educator

We urgently need to connect more kids with nature and to foster an appreciation for the foundational role that nature has in all our lives. This book is so timely as we experience the joint crises of biodiversity loss and climate change. By providing children with age-appropriate stewardship and nature experiences, we are more likely to raise children who care for the Earth. Encouraging children of all ages to have a deep bond with nature helps to shape the kind of informed, compassionate leaders that the future so desperately needs.

— John Hassell, Director of Communications and Engagement, Ontario Nature

The Wild Path Home offers a beautiful resource to help embed outdoor and environmental knowledge and care for the world around us, all connected to the natural development of young people as they go through education of all types. The age Landmarks help tie education opportunities to youth developmental characteristics from birth to age 17 in clear and helpful ways. Clearly written with Why and How sections, and highlighted with stunning photographs of the learning in action, this book is a delight to use and easy to implement for educators of children of all ages.

— Jade Berrill, Director of Learning, The Outdoor Learning School & Store

The Wild Path Home is inspired reading for educators and families alike. It provides those working with young people with a foundation, knowledge, and skills to develop children's capacities for hope, kinship, and action in a changing world. Full of age-appropriate activity ideas, suggested resources, and photographs that beautifully illustrate key concepts, this book will galvanize individuals and communities to lead learning with the next generation that addresses local and global issues in impactful ways.

— Dr. Hilary Inwood, Dept. of Curriculum, Teaching & Learning;
Coordinator, Sustainability & Climate Action Network;
Ontario Institute for Studies in Education, University of Toronto

In an age of excessive screen time, growing social disconnection, and community division, *The Wild Path Home* offers a much-needed recipe for raising nature-loving, healthy children—kids who will grow into compassionate stewards of our natural environment, and sow seeds of love and kindness within their communities. I will eagerly use it with my own young kids, and in my work as a public health leader, advocating for solutions to health challenges that ground ourselves in our shared connection to nature and to each other.

— Dr Thomas Piggott MD Ph.D., Medical Officer of Health;
CEO, Peterborough Public Health; nature lover and Dad

The *Wild Path Home* is just what we've been looking for. It is an inspiring and timely resource that reimagines the purpose of education in an era of ecological and social uncertainty. Rooted in principles of relationality, justice, and care, this book offers a powerful framework for parents, community leaders, and educators seeking to engage children in meaningful, transformative learning. It not only highlights the importance of land-based and place-conscious education, but also models how schools, homes, and community buildings can become sites of hope, resistance, and regeneration. I highly recommend this resource to anyone committed to teaching for a more just and sustainable future.

— Anne Corkery, Assistant Professor, School of Education, Trent University

There has never been a more critical moment for community connection—with each other, with our more-than-human relations, and with our home, the natural world. This vital community collaborative is both the invitation AND the map along the path to an enduring connection—a wild way through the compounding and deeply interwoven crises we collectively face, both social and environmental. This guide affords inspiration, agency, and a critical next step. Through fun and authentic connection, this wild path invites us into the outdoors every day, in all seasons. We can embark on this wild path home at every age and stage, letting our children lead the way.

— Karen O'Krafka, Environmental Educator and Past-president,
Council of Outdoor Educators of Ontario

The Wild Path Home digs deep into the research and understanding that environmental educators and parents have known for years. Spending time in nature and developing respect for the natural world leads to a closer relationship with nature, which in turn increases caring attitudes and a desire to take action to help our planet stay healthy. This book will be a valuable guide for educators and environmental mentors who hope to practise and instill kinship with the Earth. In the words of Jacob Rodenburg, "If ever there is an ethic we can teach our children, it's the idea that both nature and people can share the same space, and both can thrive."

— Judy Halpern, Consultant, Learning for a Sustainable Future

As an elementary teacher who is passionate about environmental education and feels a strong sense of responsibility to nurture a nature connection in my students, I have long wished for a resource like this. It not only outlines what I need to support and justify land-based education, but also acts as a practical guide that is tied to age and stage, so that I can use it in tandem with my curriculum as I plan learning experiences. When I am choosing how to approach environmental issues like climate change with young people, I worry about inspiring anxiety because I know that they are deeply concerned about the welfare of our planet. This guide is not only a source of practical action that I can take to counter my student's fear, but also of hope as it focuses on inspiring a sense of wonder regarding our natural world. I have already used the Pathways to Stewardship and Kinship landmarks in my teaching for a number of years and look forward to having this guide as an accessible personal reference. I am also excited to offer it to my school board leadership as a planning tool for curriculum, policy making, and professional development for educators.

— Sheila Potter, educator and naturalist

Any educator will find this book a valuable resource. *The Wild Path Home* is an inspiring guide to how children can become land-stewards through age-appropriate activities.

— Mikaela Cannon, outdoor educator and
author of *Foraging as a Way of Life*

THE WILD PATH HOME

THE
WILD PATH
HOME

A GUIDE TO
RAISING THE
EARTH STEWARDS OF TOMORROW

JACOB RODENBURG · CATHY DUECK

new society
PUBLISHERS

Cover design by Diane McIntosh.
Cover image: ©iStock

Printed in Canada. First printing August, 2025.

Inquiries regarding requests to reprint all or part of *The Wild Path Home*
should be addressed to New Society Publishers at the address below.
To order directly from the publishers, please call 250-247-9737
or order online at www.newsociety.com.

Any other inquiries can be directed by mail to:

New Society Publishers
P.O. Box 189, Gabriola Island, BC V0R 1X0, Canada
(250) 247-9737

New Society Publishers is EU Compliant. See newsociety.com for more information.

LIBRARY AND ARCHIVES CANADA CATALOGUING IN PUBLICATION

Title: The wild path home : a guide to raising the Earth stewards of tomorrow /
Jacob Rodenburg, Cathy Dueck.

Names: Rodenburg, Jacob, 1960– author | Dueck, Cathy, author.

Description: Includes bibliographical references and index.

Identifiers: Canadiana (print) 20250181630 | Canadiana (ebook) 20250181789 |
ISBN 9781774060148 (softcover) | ISBN 9781550928075 (PDF) | ISBN 9781771424035 (EPUB)

Subjects: LCSH: Child rearing—Environmental aspects. | LCSH: Sustainable living. |
LCSH: Environmentalism. | LCSH: Environmental health.

Classification: LCC HQ755.8 .R63 2025 | DDC 649.1—dc23

Funded by the Government of Canada | Financé par le gouvernement du Canada

Canadä

New Society Publishers' mission is to publish books that contribute in fundamental ways
to building an ecologically sustainable and just society,
and to do so with the least possible impact on the environment,
in a manner that models this vision.

MIX
Paper | Supporting responsible forestry
FSC
www.fsc.org FSC® C016245

EU COMPLIANCE PARTNER

Certified B Corporation

new society
PUBLISHERS

Contents

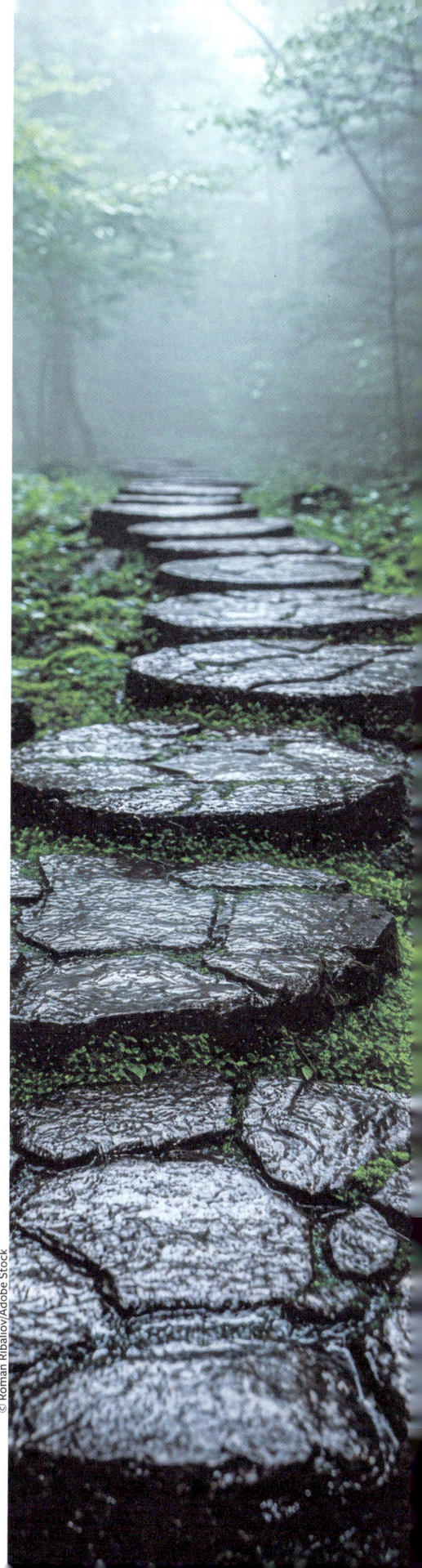

© Roman Ribaliov/Adobe Stock

Acknowledgments

The authors would like to thank the many people in Peterborough/Nogojiwanong who helped to develop, support, and pilot-test the Pathway Project over eight years. There are far too many to name individually, but we are grateful to each and every one of you.

Of particular note is our colleague Anne-Marie Jackson, who worked for several years as our Communications Coordinator, and who developed an excellent series of how-to videos which are showcased on the Pathway website (pathwayproject.ca).

A special thanks also to Nicole Bell, Anishinaabekwe, who was our key Indigenous advisor throughout the development of the project, and from whom we all learned so much.

We are also grateful to the Ontario Trillium Foundation for seeing great value in this work, and who enabled so many young people to participate through their generous funding.

A big hug to all the amazing educators who work so hard every day to bring the wonders of nature into the lives of their students, and who are such wonderful role models of kindness and compassion in a world that sometimes seems topsy-turvy.

It's been our great privilege to work with so many dedicated people from a great variety of community organizations. Camp Kawartha stepped up to act as the lead, taking financial and legal responsibility while sharing so much brilliant experience from their talented staff. Other groups whose staff shared their time and talents include Otonabee Conservation, the Kawartha Pine Ridge District School Board, the Peterborough Victoria Northumberland and Clarington Catholic District School Board, GreenUP, Peterborough Public Health, Trent University, Fleming College, Peterborough Child and Family Centres, Think Outside, Investing in Quality Peterborough, City of Peterborough, Riverview Park and Zoo, and the Kawartha World Issues Centre.

And to all the wonderful young people in our community and throughout the world who are becoming the Earth Stewards of tomorrow: you inspire us and light the fires of hope, reminding us of the joy of sharing this miraculous world together.

Introduction

Imagine...

Imagine a world where all children grow up respecting themselves, each other, and the natural world around them. Imagine a society where health is the driving force—mental health, physical health, and environmental health—shaping the way kids are raised and guiding the priorities of our communities. Imagine leaders from around the world making environmental literacy a key focus for all education, inspiring a deep sense of caring and belonging in children and youth.

Is this an impossible dream? The Pathway to Stewardship and Kinship is a program that offers a road map to making

Credit: Matthew Walmsley

it a reality for every child, everywhere. It's a starting point for anyone involved in a child's life—parents, teachers, caregivers, grandparents, relatives, and community leaders. In a world consumed by economics and productivity, this approach shifts the focus to nurturing relationships—with each other and with the lands and waters that sustain us. It's the first step toward a future that is equitable, sustainable, and healthy— a journey guided by Indigenous wisdom.

Why Now?

We live in a time of deep division. Political forces have never been more polarized and antagonistic. The pursuit of power, along with religious and racial tensions, have fueled wars of devastating intensity. The gap between rich and poor continues to widen, leaving many more families around the world in the grip of poverty. Leaders often spread misinformation, and the virtual world of digital technology blurs the line between truth and fiction. The COVID-19 pandemic that started in 2020 amplified these trends, enforcing social separation for three years and increasing our reliance on virtual communication. This stress and frustration sparked a surge of vaccine conspiracy theories and protests against public health measures.

Where are our children in this whirlpool of challenges? How do they deal with an increasingly complex and technologically saturated world? Psychologists are sounding the alarm over an epidemic of mental health issues—ranging from anxiety disorders in children and rising rates of suicide and substance abuse among youth to widespread feelings of alienation and depression. Teachers report an increase in special needs, including attention deficits and antisocial behaviors, a trend that has worsened since the social isolation of the pandemic. Health care providers are also concerned about declining fitness and health, driven by too little physical activity and too much screen time. This sedentary indoor lifestyle is contributing to a host of health challenges, including heart disease and diabetes.

These issues have led to a surge in research investigating causes and potential solutions, starting in early childhood. With the rise of instant worldwide communication through the internet, stories of child abduction and abuse have led to an exaggerated sense of danger in the real world, even though rates of crime, violence, and other physical risks to children have continued to drop since the 1990s.[1] Parents today have grown increasingly protective, often limiting activities that were once a regular part of childhood, like walking to school, unsupervised outdoor play, tree climbing, and other forms of exploration that might lead to minor scrapes and bruises. In response, many schools have also adopted a more cautious approach, curbing field trips and restricting access to ponds, trees, and natural meadows—all in an effort to avoid potential legal risks.

As a result, children are losing crucial opportunities to test their boundaries, develop resilience, and learn how to assess risk and solve real-world problems independently. Meanwhile, early and often unsupervised exposure to screens, the internet, and social media contributes to social anxiety, cyberbullying, and exposure to online predators and violence. While parents believe keeping their children indoors offers greater safety, the reality is often the opposite. This contradiction is known as the "protection paradox."[2]

Can we restore a sense of security, belonging, and confidence to those families and communities as they struggle to raise children today? Research consistently points to a common solution: children need ample time for unstructured play in natural outdoor settings, where they can take reasonable risks, solve real-life challenges, engage in creative play, and experience the awe and wonder of the world around them.

Benefits of Unstructured Outdoor Play

- **Better physical health:** children engage in more active play when they're outdoors, resulting in better cardiorespiratory health and improved overall physical fitness.[3,4,5]
- **Better mental health:** numerous studies point to reduced stress, anxiety, and depression when children and youth have plenty of unstructured time in diverse natural environments.[6,7,8,9]
- **Improves concentration:** especially for children with attention deficits and hyperactivity, outdoor time improves the ability to focus attention.[10,11]
- **Develops cooperation, collaboration, and self-regulation:** children's ability to overcome obstacles cooperatively and manage their impulses improves when they engage in creative, unstructured outdoor play.[12,13,14]
- **Stimulates creativity:** in natural environments, access to a wide range of materials gives the imagination free rein. Branches can become a fort, twigs and stones can become a house for a fairy, leaves and pebbles of many colors can become beautiful ephemeral art. The possibilities are endless.[15,16]
- **Enhances self-esteem:** learning to manage moderate risks (jumping from rock to rock, climbing trees, etc.) helps children overcome fears and build confidence in their abilities.[17,18]
- **Develops problem-solving and cognitive abilities:** in outdoor environments with so many variables, children exercise their senses as well as their bodies, which in turn stimulates the brain. Studies show

Credit: Heather Snowball

improvements in cognitive performance in schools with regular opportunities to interact with the natural environment.[19, 20, 21]

• **Sparks a lifelong interest in learning:** when children are less stressed, in better physical and mental health, and better able to cope with challenges, they bring a more positive attitude to learning. When learning is infused with a spirit of excitement and discovery, a child's innate curiosity is sparked, and a passion for learning can flourish.[22, 23]

In Waldorf preschool programs, the toys are derived from the natural world, offering limitless possibilities for imaginative and creative play. "Bark from trees can be made into bark boats that float in a puddle. Pieces of moss can be gathered and used to cover the floor of homes that children build in the hollows of tree roots…. Colored leaves in autumn and dandelions in the spring are woven into crowns that are worn by children." [24]

Mentorship

To nurture young people with a lasting sense of connection to and care for their world, plenty of time to play in nature is the first essential ingredient. Studies highlight a second key factor: having a caring role model or mentor who offers the freedom to explore and discover, while also sharing a sense of awe and wonder, and fostering empathy and respect. These studies explored the links between childhood experiences and interest in protecting the environment in later years.[25, 26, 27]

For very young children, mentors are typically parents, close relatives, or caregivers. As children enter elementary school, mentors may expand to include teachers, older students, or youth leaders. For teens, mentors often include older peers, secondary school teachers, or community leaders. Effective mentors allow young people to guide their own learning, offering support rather than taking the lead. Louise Chawla identifies four key qualities of a strong environmental role model:

Credit: Madeleine Endicott

- **Care for the land:** through actions or words, encourages care for the land as important for personal identity and well-being;
- **Disapproval of destructive practices:** pointing out to a child when activities are hurtful or harmful to the land or its inhabitants;
- **Pleasure at being out in nature:** demonstrating enjoyment of being in the natural world;
- **Fascination with elements of the natural world:** sharing interesting observations of the Earth, water, sky, plants, and animals (here, a sense of wonder is more important than instructing with facts).[28]

> Only in a natural area children are discovering the qualities of the world with which humanity evolved, on which human existence depends. As children play and explore in nature, they are becoming familiar with essential properties of the biosphere.
>
> — Louise Chawla[29]

Giving children many opportunities to have positive experiences in the natural world, with access to a caring mentor, is a powerful way to stimulate a sense of community, of belonging, and a sense of responsibility toward the world around them.[30] By encouraging children to engage in simple acts of stewardship, we inspire advocacy and an ethic of caring. Every young person, at any age, can have positive impacts on the environment. It's a simple yet powerful formula for nurturing the responsible, engaged citizens of the future.

Sadly, most childhood experiences in our modern world push children in exactly the opposite direction. Many children's days are prescheduled from the moment they wake. Fascination with technology leads to children spending an average of 7½ hours a day in front of a screen.[31] The pressure to succeed and secure a good job often translates into even more sedentary indoor time spent focusing on literacy and numeracy skills, imagined to be the golden key to future success.

The Pathway to Stewardship and Kinship offers a straightforward approach

Credit: Heather Snowball

Credit: Geri-Lynn Cajindos

to giving children what they truly need to thrive—time to connect with nature and their community, a foundation that helps them grow into their healthiest, most resilient selves, eager to face the future and its challenges.

What Is Stewardship?

Words can carry many meanings. When we talk about stewardship, we refer to a deep sense of connection, care, and responsibility for one another and the natural world that sustains us. Stewardship involves personal actions aimed at protecting and improving the health and well-being of both human and natural communities, which are intricately interconnected. It also acknowledges that human health and survival are entirely dependent on thriving ecosystems.

True stewardship means living in ways that respect and strengthen the interdependent web of life that connects all of us. As John Muir wisely observed, "When we try to pick out anything by itself, we find it hitched to everything else in the universe."[32]

What Is Kinship?

From a First Peoples' perspective, we are kin to the living and nonliving world around us. The air, water, soil, rocks, plants, insects, and animals are all part of our community. The term "kinship" recognizes that we belong to the fascinating and complex web of life, and that each strand in the web is equally important. It is crucial to respect the right of other species to exist, acknowledge their place in the world, and be ready to learn from them.

Fostering stewardship and kinship is a proactive undertaking that must involve many caring mentors—educators, parents, relatives, and youth leaders. These mentors provide a community network that can encourage discovery, share a sense of awe and wonder, and actively cultivate empathy and respect for all life. As children begin to learn how the world functions, they understand the impacts that people can have and explore solutions to challenges in their community. As young people continue to grow toward their teen years, community mentors help to guide them in developing leadership skills by participating in local action, encouraging confidence, agency, hope, and belonging.

The Road Toward Stewardship and Kinship

The Pathway to Stewardship and Kinship is a simple, stepwise plan developed, field-tested, and implemented by a community of people who care about kids—perhaps a community just like yours! We believe that cultivating stewardship and kinship in all children is at least as important as literacy and numeracy, and deserves a carefully developed strategy that embraces every young person, throughout their growth and development.

Developing the Pathway involved extensive research and collaboration with people from all walks of life—parents, teachers, early childhood educators, Indigenous Elders, psychologists, health professionals, politicians, administrators, environmentalists, artists, and students. Community interviews were followed by reviewing research into child development, health, psychology, and environmental education.[33] The results were distilled into a series of 30 simple but powerful "Landmark" activities.

> **land·mark** 1. An object or feature of a landscape or town that is easily seen from a distance, especially one that enables someone to establish their location; 2. An event, discovery or change marking an important stage or turning point in something.
> — *Oxford Dictionary*

Each Landmark links to a particular stage of child development, opening the doors to strong lifelong relationships with the land and all its inhabitants, and a sense of responsibility for its well-being.

The Landmarks also provide ideas for inspiring action in a way that builds hope and empowerment, instead of fostering fear. Nurturing stewardship and deepening relationships means not only educating the *head* but also providing experiences that cultivate the *heart* and translate into positive action for the *hands*.[34] When these are supported by a host of strong relationships, the resulting rich tapestry connects us to the world around us, giving a sense of belonging, of home, and our place in it. This is what anchors us to our world.

Indigenous Wisdom

As we continue to learn how to make wise parenting, education, and lifestyle choices, we are grateful for the wisdom of Indigenous cultures that have lived and thrived in a reciprocal and sustainable relationship with the land for millennia. While every Indigenous culture is a unique expression of its relationship with the land, there are common threads that provide a beacon of direction for all of us in this modern, fragmented, and ailing world.

Anishinaabe scholar and educator Nicole Bell speaks of the 4Rs as a framework for developing our relationship with the land.[35] Using the medicine wheel as a model of ongoing learning and growth, she explains that we must begin with respect (in the east, as the rising sun)—respect for ourselves and all creation. With that sense of respect, we can begin to develop meaningful relationships with each other and other beings (moving to the south). We must also bring a sense of reciprocity into all of these relationships; not only taking but considering how we can give back as well (in the west, on the medicine wheel). Finally, respect, relationship, and reciprocity grow into a sense of responsibility to live in a way that benefits all creation (moving to the north, to complete the cycle). This is an ongoing process that keeps repeating and strengthening

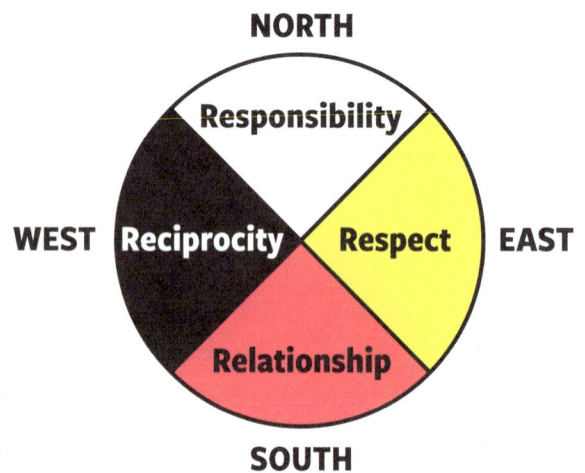

Credit: Lisa Gutoskie

Medicine wheel interpretation based on the teachings of Nicole Bell.

throughout life. These are important teachings that can provide a firm foundation for guiding children and youth as they learn, grow, and find their own unique place in the world.

Another principle common to many Indigenous cultures is the importance of intergenerational relationships in a child's education. At home, at school, and in the broader community, children benefit from having mentors of many ages, from a variety of walks of life. If these mentors honor and respect each child as an individual, they can contribute to a deep sense of belonging, of a community safety and support network.

Credit: Julie Stoter

With the support of many wise and caring mentors, and plentiful opportunities for unstructured outdoor time, the foundation is laid for a child to develop a strong sense of stewardship and kinship. We have heard from many Indigenous advisors, "The Land is the first teacher. The Land is the best teacher."

Especially in these bewildering times, this is a powerful place to begin—whoever you are, and wherever you live.

We are grateful to the many Indigenous Elders, Knowledge holders, friends, and community leaders who provided guidance in developing the Pathway to Stewardship and Kinship. Their voices can be heard throughout these pages. Our deepest thanks for your generosity.

Climate Change

During the years of developing and field-testing the Pathway to Stewardship and Kinship, climate change has become an ever more urgent issue around the world. More than ever, we need to cultivate leadership and problem-solving skills in young people of all ages, involve them in meaningful action projects, and give them opportunities to witness real-life success stories. Gloom-and-doom scenarios may boost ratings and social media hits, but they do not generate hope, nor provide incentive for long-term action. The Pathway Landmarks include many activities that guide us away from energy-consumptive activities and emphasize active trans-portation, renewable energy, and getting involved in local action projects. As young people learn how our choices impact the world around us and explore more sustainable lifestyles, their parents and other educators can benefit and learn too.

Recent research suggests that involving young people in social justice issues can be another powerful springboard toward a commitment to climate change action in adults. This is a reminder that strong social ties paired with strong environmental ties can be a catalyst for fostering personal, environmental, and community health.[36]

Who Is This For?

This stewardship framework enhances and complements existing parenting guides, school curricula, and community programs. The recommendations can be adapted to suit any school, early childhood learning center, community group, or age group, offering ideas for family activities, action projects for children of all ages, and guidance for community groups planning or running programs for youth. It creates a shared focus that benefits everyone involved.

The more widely these core principles are embraced across all sectors of a community, the greater the impact on children's health and long-term well-being—on a global scale. We don't need large budgets or elaborate preparation. We can, and must, start now.

Key Stewardship Principles

Though many activities in this book are tailored to specific ages and stages of childhood, the guiding principles are timeless, universal, and relevant for all ages, regions, and cultures. These principles form the core of the Pathway to Stewardship and Kinship's approach, providing a foundation that nurtures deep connections, a sense of responsibility, and a pathway toward building healthy, thriving communities—both now and for future generations.

Adults are powerful role models for young people. You don't need to know the names or how everything functions in nature to be an effective mentor, but you do need to bring a positive and respectful attitude. This sends a message that resonates with children long after the experience. Even very young children can be taught to touch gently, step carefully, and observe quietly, ensuring no harm is done. Always return creatures to their habitats after observing them, sharing the excitement of how amazing they are. When adults model these behaviors, the message of stewardship becomes far more impactful than words alone.

This is a summary of the stewardship principles that are equally important for all ages and should be practiced and en-couraged by everyone who interacts with young people.

Respect for Each Other and All Nature

A fundamental value in fostering stewardship is the recognition that life, in all its forms, is amazing and that every being—human or nonhuman—deserves respect. As adults, it may require some personal reflection and the resolve to overcome fears or biases we've developed, so that we can become effective mentors for children on their journey toward stewardship. One simple way to cultivate respect and empathy is by providing children with positive experiences with animals, whether wild or domestic.

Credit: Jessica Carey

In Indigenous worldviews, even non-living elements are believed to possess a spirit—threads that weave together the fabric of life. Water, air, rocks, along with plants and animals, all play vital roles in creation and are equally deserving of respectful relationships.

This principle is so fundamental, it is recognized in the United Nations Convention on the Rights of the Child that "education of the child shall be directed to the development of respect for the natural environment."[37]

Credit: Sonya Friesen

Sense of Awe and Wonder

We are never too old to appreciate the wonders of life, or to encourage and share that sense of awe that results from truly seeing the world around us. Awe and wonder fuel curiosity and a desire to learn, sparking the lifelong joy of discovery. We forget as adults how powerful language can be. If you want to cultivate a sense of wonder, you need to use the language of wonder. Phrases such as "Isn't this amazing! I wonder…let's go find out more" help to open a child's eyes to the miracle of life itself.[38,39]

When we truly exercise all our senses regularly, and notice tiny things as well as the big picture, we're much more likely to see the miracles happening around us every day. It's remarkable how much more you see when you travel on foot, rather than any type of faster moving vehicle. Slowing down, looking carefully, listening intently, smelling the seasonal changes, and encouraging children to touch the textures of nature draw us into a fascinating community of life that too often goes unnoticed.

Natural Curiosity and Discovery

It is a wise teacher who knows when to share information and when to step back and let natural discoveries take place. Overloading children with facts can inhibit their interest in learning and discovering. The most effective learning centers on the child, not the teacher. Keep in mind that the engine of learning is curiosity.[40] As adults, we need to remember that a name

Credit: Danielle Blondin

Credit: Grace Beaumont

or a label is merely a beginning point. It is the start of a story—an intriguing one—and it is up to you to keep the story going! A good question should invite other questions. Think about your questions as ways to encourage kids to ask why, to wonder, to marvel at the natural world and to promote further exploration.

Sense of Place

An important part of developing a sense of security and belonging is spending enough time outdoors in the same place to become deeply familiar and connected with it. For those who have developed a particular attachment to a place when growing up, that sense of place becomes part of their identity. It is important to give children plenty of time to develop those deep attachments to place, whether that's a favorite park, a cottage, a camp, or other outdoor place with special memories.[41]

Every place on Earth is unique. Is your home near a forest, a grassland, a wetland or lake? Do you live on the arctic tundra, near a desert, or on the ocean shore? Is your home in a tropical or temperate rainforest, or do you live in the mountains? Getting to know and fall in love with the place you call home kindles a sense of family and belonging that extends beyond your human relatives. No other person on Earth is the same as you, and the place you live is unlike anywhere else. Taking the time to explore and grow "roots" in your home place takes time and patience but enriches our lives profoundly with a deep and lasting sense of connection.

Credit: Kathy Connelly

Sense of Gratitude

Often, we become so preoccupied by daily pressures and challenges that we become oblivious to the many wonders and gifts around us. Wisdom from many faith communities and Indigenous ceremonies remind us to take time every day to recognize and appreciate the many gifts of creation.[42] This can begin with being more aware and mindful of the world around us—taking time to be calm, quiet, and present in the moment and seeing the wonders around us. It is also important to practice gratitude. The act of giving thanks helps to strengthen our connections to each other and to the special places that are an integral part of every community. A sense of gratitude helps to strengthen mental health at all stages of life.

> The container is now full and the class is very excited to begin boiling the sap tomorrow. Wow! We are thankful for the maples trees that gave us this gift.
>
> — Indrani Talapatra (teacher)

Interconnectedness

Children benefit from many opportunities to learn how our lives are connected to the lives of other people and all things in nature. We share the same air, the same water—the food we eat contains nutrients made from atoms that have circulated for eons.[43] This understanding reinforces our innate need to belong. Stewardship involves understanding that we belong to a community that extends far beyond our close friends and relatives and includes the living systems that are integral to health. Let's encourage all children to get to know and love their "Neighbourwood."[44]

When you observe something in nature, think about the many ways that it is connected to its environment. What kind of habitat does it live in? What does it eat? Who eats it? How does it reproduce? How much space does it need? How does it move? Where does it find water? What was it doing when you discovered it? Was it camouflaged, or did it stand out from the things around it? Are there many others like it, or is this the only one? The questions we can ask to deepen our understanding of the world around us are limitless.

While the science of ecology is a relatively new field of study, Indigenous Peoples have known for millennia that all things, both living and nonliving, are deeply interconnected and equally important. Humans are no more important than any other part of creation, with each being a vital thread in the tapestry of life.

Mentorship at All Ages

A recurring theme in many research studies is the importance of children spending time with a caring mentor in developing a sense of caring and connection that will last into their adult lives. In the early years, the mentor is usually a close relative such

Credit: Indrani Talapatra

Credit: Lisa Gutoskie

Credit: Nancy Hurley

Credit: Karen Brown

as a parent or grandparent who spends time with the child, exploring together and sharing the delights of discovery.[45]

As a child grows older, mentors begin to include favorite teachers or other youth leaders who become trusted and admired role models.[46] While having knowledge to share with a child is helpful, it's more important to share an interest and inspire curiosity.[47] Seniors can be valuable mentors for children, and opportunities for intergenerational learning can be of mutual benefit.

While access to many mentors is important, children also need plenty of time to explore on their own and with other children. Wise mentors leave plenty of room for self-guided discovery. A mentor provides support and encouragement, but minimal "instruction." It is also important to remember that older children can become mentors of younger children, and that the art of mentoring is a skill that can be practiced and refined over time.

Learning from Local Indigenous Cultures

Wherever you live around the world, local Indigenous cultures provide invaluable guidance in learning to live in harmony with the unique place that is your home. While Euro-Western economic systems put a stamp of "sameness" on everything from buildings, services, and even to landscapes (e.g., the ubiquitous lawn aesthetic), Indigenous cultures developed over millennia as a result of intimate relationships specifically with the land where they lived.

Look for opportunities to learn from your region's Indigenous Peoples at special events throughout the year, and by getting to know, support, and develop friendships with First Peoples wherever you live. This is a meaningful way to help restore the value and significance of peoples and cultures that have been marginalized for far too long. Indigenous knowledge offers essential insights in our journey to relearn how to live sustainably in harmony with the land—no matter where we call home.

Overcoming Fears

Not everyone feels comfortable and safe outdoors—especially in natural areas. Building comfort and security outdoors is something you can learn, with patience and practice. This can begin by dressing appropriately for all weather, so you feel comfortable in all conditions. If possible, provide very young children with a mud suit to wear over their clothes so they can delight in jumping in puddles and playing in mud without a fear of getting dirty. Dress in layers in cold or inclement weather so it's easy to remove layers if necessary to stay comfortable. Get children used to wearing hats and sunscreen on sunny summer days and pay attention to public health recommendations for your area (e.g., wearing long pants and tucking in socks in places where ticks are present and doing tick checks when you get home). Don't let fear of harm keep you and your children indoors! The more you become familiar with the place where you live, and make reasonable precautions a part of daily

Credit: Anne-Marie Jackson

routines, the more comfortable everyone will be spending time outdoors.

Learn the real dangers (falling over cliffs, drowning, etc.) and learn how to handle them (stay away from the edge, learn how to swim, etc.). Fear of the dark, animals, insects, snakes, etc. can all be overcome with patience. Working on our own fears as adults can help our children become more confident and less anxious and fearful. Never push a fearful person into doing something that makes them uncomfortable. Rather encourage them to take small, incremental steps in managing their fear, with plenty of praise and support for each step along the way.

Child development experts are concerned that overprotecting our kids, with the best of intentions, has the opposite result to what we wish for them. When

Credit: Craig Brant

children aren't given the opportunity to engage in moderately risky play, they don't develop confidence in their abilities and are more likely to remain anxious and fearful. When a child, for example, masters the skill of crossing a small stream by balancing on a log, they are proud of their accomplishment and eager to share their mastery with others. Getting scrapes is a normal part of growing up, and letting children take reasonable risks (jumping, climbing) helps them to stretch their abilities, learn their limits, and build resiliency. Learning to overcome fears literally opens the doors to a world of wonders.[48, 49, 50]

Practicing Independence

Another fundamental principle confirmed by many research studies is the importance of giving children opportunities to practice as much independence as their age and capabilities will allow. That can begin at a very young age by encouraging children's innate desire to help. Letting young children set the table or assist with simple food preparation helps to develop skills for independence and fosters a sense of self-worth.[51]

Jonathan Haidt, author of *The Anxious Generation*, recounts the story of a seventh grade teacher who was so concerned about her students' fears of doing anything on their own (for risk of failure or ridicule) that she created a new program that she called the Let Grow Project. Over the school year, each student was given a homework assignment to complete 20 different tasks on their own, and report on their experiences. Suggested tasks included things such as doing the family laundry, preparing a simple meal, walking to the park with friends, riding the bus without an adult, or other ideas of their own choosing. Students were so excited by their accomplishments, there was a noticeable boost in confidence and self-esteem.[52] Parents were amazed by the capabilities and responsibility shown by their children and felt comfortable giving them more freedom.

Since the rise of overprotective parenting in the 1980s and 1990s, parents have been afraid to let their children out of their sight. That has resulted in driving children to school instead of letting them walk, driving them to the local store for picking up a few groceries, and accompanying them on all outdoor activities. This has become such a cultural norm that parents have actually been reprimanded or charged for letting their children walk or cycle in their neighborhood on their own or to ride the bus to visit a friend or attend a program. The days of neighborhood kids playing outside without an adult are a thing of the past in many communities. This has further restricted physical activity and limited children's opportunities to make decisions, travel and play safely on their own, resolve conflicts with their friends, and look after each other without an adult's presence.

Travelling without an adult is what the experts call "independent mobility." Travelling a regular route alone or with peers is an excellent way to sharpen skills of observation, enhance interest in and knowledge of the environment, as well as

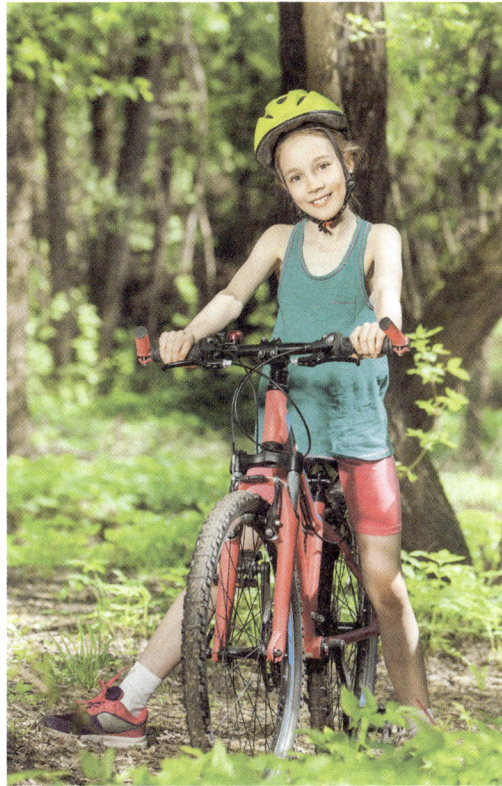

Credit: iStock

promote a sense of security.[53,54] Children permitted to travel regular routes independently also tend to play outside more frequently, leading to the host of benefits previously mentioned, such as stress reduction, creative thinking, problem-solving, and increased self-esteem.[55]

While very young children cannot be expected to understand the dangers of traffic, older children who demonstrate awareness, understanding, and basic safety should be rewarded with corresponding degrees of freedom of movement. This is a difficult concept for today's cautious parents to implement for fear of reprisal from other adults. It can be helpful to work together with other local parents to encourage children walking or cycling together to school, a local park, or a nearby convenience store. The 2015 report from Participaction, called *The Biggest Risk Is Keeping Kids Indoors*, is a well-documented examination of the risks and benefits of independent mobility.[56] The age at which children can handle independent travel will vary from child to child, but typically between ages 8 and 10 is a good time to start the process with short-distance travel, gradually broadening the range as young people demonstrate responsibility and the ability to handle the freedom.

Accessibility

Differences in physical abilities need not deter anyone from enjoying time in nature. Tools such as all-terrain wheelchairs may be available to borrow from a variety of nearby children's service providers. Many outdoor centers and public parks now provide tools for accessibility, so that anyone can participate in nature adventures. These can include tools such as hoists for lifting children into boats and special harnesses for wall climbing. Exciting new developments in accessibility are becoming rapidly available to everyone, so be sure to check out what is available in your area.

Limits to Screen Time

Another recurring recommendation, both in research and feedback from parents and community leaders, points to the benefits of limiting screen time—television, computer, and especially smartphones. Too much screen time severely limits physical activity, social and creative development, as well as causing a sense of separation from the real world. While technology can offer many benefits, too much of it can be toxic to healthy development.[57] Many educators are so concerned about the negative impacts of social media on young people's mental health that they propose banning cell phones from schools and severely restricting young people's access to smartphones altogether.

Jonathan Haidt has researched the impact of smartphones and 24/7 access to social media on the mental health of young people in his 2024 book *The Anxious Generation: How the Great Rewiring of Childhood Is Causing an Epidemic of Mental Illness*. He points out dramatic numbers of anxiety and depression disorders, particularly in preteen and teenaged girls, beginning around 2010. This was the time when smartphones became widely available, and most teens had their own smartphone within a couple of years. Girls became obsessed with social media, which started a catastrophic rise in anxiety. Boys tended to be more obsessed by video gaming and access to pornography. In both cases, in-person social interaction declined significantly, replaced by online social interaction. Depressive disorders in the U.S. rose by 145 percent in teenaged girls between 2010 and 2020 and by 161 percent in teenaged boys. Preteen suicide rates rose sharply during the same time period.

Similar trends were also found in Canada, Australia, England, and the Nordic countries. Haidt recommends that schools be phone-free for students, and that children should not have access to smartphones before high school, with no exposure to social media before age 16. He further recommends that parents oversee their children's online time and implement parental controls and content filters on all digital devices at home. Numerous health agencies have similar recommendations on limiting children's access to recreational screen time: none before age 2 years; limited access between 2 and 6 years; no more than two hours per day for 6- to 12-year-olds. For teens, it's helpful to discuss and agree on family guidelines together, such as no screens during mealtimes and everyone turning off screens an hour before bedtime.

However, there can be positive use made of tablets and smartphones in outdoor education by harnessing young people's interest in technology. Some apps are helpful in enhancing outdoor knowledge and skill and, if used with discretion, can be powerful learning tools for youth. Apps such as iNaturalist can help with identifying natural features, and Citizen Science apps such as eBird can encourage young people to participate in collecting information about wildlife to assist with local and regional conservation efforts.

Creative Expression

The arts provide one of the most powerful ways of developing and expressing morals and values, which are a foundation of mental health. Research refers repeatedly to the importance of providing opportunities at all ages to express feelings. Discussion,

Credit: Julie Stoter

Credit: Craig Brant

painting, drama, stories, music, dance, poetry, photography, sculpture, and music are a few examples of the ways we express ourselves and learn from others.

The arts are also an important vehicle for developing empathy and sympathy by imagining ourselves inside the lives of other beings.[58] While young people may prefer certain ways of expressing themselves at various ages, creative expression overall remains an important factor in healthy development throughout life.[59] The natural world provides many opportunities for inspiration and reflection, as well as providing a wealth of materials for artistic expression.

Action and Giving Back

Everyone, no matter their age or ability, can do something positive for the world around them. Tending a garden, raising butterflies, caring for a natural area, and reducing our energy consumption are just some of the simple ways we empower young people to make a positive impact right in their own communities. A foundational stewardship principle is considering how we can give back to the Earth that supports us. Kids can solve a problem provided they are given the right tools and strategies for their age. Every positive action leads to a sense of hope, and every bit of hope is empowering.

However, in a world grappling with climate change, species loss, and pollution, it's crucial to provide children with age-appropriate information that matches their emotional and cognitive development. Introducing overwhelming problems, especially at too young an age, risks fostering what some call "ecophobia" or "ecoanxiety," a sense of paralysis or apathy

that stems from feeling powerless in the face of global environmental challenges. Instead, by focusing on solutions and actions close to home, we can help children see how their efforts matter. As kids grow older, they can begin to explore the idea of sustainable living: reducing their carbon footprint, exploring alternatives to fossil fuels, learning about product life cycle and social justice issues.[60,61,62] Taking action and finding ways to give back helps young people develop a sense of agency, empowerment, and hope for the future.

Never Too Late to Start

What if your kids are already in their middle years or teens? Is it too late to start the Pathway Landmarks and activities? It's never too late—even if you're an adult who has rarely been outdoors and has many fears. This program is for everyone, at all ages, regardless of when you start on the path. The key is to support each other as we learn new ideas and skills for responsible and rewarding involvement within our communities and our world. Each Landmark is a springboard for activities that help children to grow, thrive, and connect to the world around them.

The Power of Joy

All of the suggested activities in this book are grounded in the knowledge that great joy can be found in nature and being outdoors, and that joy is contagious. Laughter, a powerful stress reliever, and a sense of fun are woven into every Landmark. The Pathway to Stewardship and Kinship is a rewarding journey for both children and their mentors. We all benefit by raising healthy children for a healthy planet.

Credit: Shayla Bush

> We visited our outdoor learning space and explored new trails. We also noticed how we felt when we were outside!
>
> — Shayla Bush (teacher)

Using the Key Themes Chart

All of these basic principles are important throughout childhood and adulthood, as we travel together toward a culture of stewardship and kinship. The following chart summarizes samples of key stewardship themes common to all ages, as well as those that are particularly critical at specific stages of development. This is a broad overview of concepts underlying the principles, landmarks, and activities suggested in the Landmarks section that follows.

(Legend: ■ = intense color; ▨ = light shading; blank = unshaded)

Key Stewardship Themes	Age 0–3	Age 4–5	Age 6–7	Age 8–10	Age 11–13	Age 14–18
Respect for self, other people, other beings	■	■	■	■	■	■
Sense of belonging, sense of place	■	■	■	■	■	■
Positive interactions with animals	■	■	■	■	■	■
Creative expression	■	■	■	■	■	■
Empathy for others	▨	■	■	■	■	■
Sensory development	■	■	■	■	▨	▨
Learning through play	■	■	■	■	■	▨
Learning through social interaction	▨	■	■	■	■	■
Outdoor exploration with a supportive relative	■	■	■	■	▨	
Having a supportive, knowledgeable mentor (other than family)		▨	■	■	■	■
Understanding interconnectedness of all things		▨	▨	■	■	■
Participation in hands-on community projects			▨	■	■	■
Opportunities to practice leadership			▨	▨	■	■
Unguided outdoor exploration			■	■	■	■
Longer periods outdoors (overnight camping, hiking, canoeing)					▨	■
More complex outdoor skills and adventure sports						■
Exploring social justice and other global/community issues				▨	■	■

Note:

• Each theme is accompanied by a colored bar showing ideal ages for exploring various skills and concepts.

• Remember that everyone is different, and we all grow and develop in different ways! These themes are a simple visual guide only.

• Light shading in a bar indicates some benefit for that age group, deepening to more intense color as potential benefits increase at other stages of life.

Landmarks

Credit: Jackie Donaldson

We can only see a short distance ahead, but we can see plenty there that needs to be done.
— ALAN TURING (mathematician)

About the Landmarks

In a world facing ever-growing environmental and social challenges, we have crafted a road map to guide children on their journey to becoming compassionate stewards of the Earth. The Pathway to Stewardship and Kinship's approach is grounded in research, shaped by community input, and has been thoroughly tested across an entire region. This strategic framework offers critical experiences tailored to each stage of a child's development, nurturing a sense of caring, connection, and responsibility toward one another and the Earth we share. These experiences are distilled into the Pathway's 30 Landmarks—a simple yet powerful tool for families, educators, and communities to engage with young people at home, in school, and beyond. Each Landmark represents a collection of age-appropriate actions designed to inspire, activate, and empower children and youth to create a better world for themselves and future generations.

The Pathway features several Landmarks for each age group, building on the skills and experiences gained from earlier stages. However, the journey is not strictly linear. Older teens can benefit from revisiting earlier Landmarks, while younger children may find inspiration in Landmarks slightly ahead of their age group. Spanning the years from birth through late adolescence, these Landmarks can be easily integrated into any program. No matter your experience or background, the Landmarks offer a common starting point for fostering stewardship and kinship. There are many ways to interpret and implement each Landmark. We've suggested a few ways that we have found children have responded to but you may have your own ideas, and that's excellent! Adapt these ideas to fit your interests, abilities, and resources but keep in mind—repeated involvement is the key.

The following pages illustrate how each Landmark can be brought to life, showcasing how communities can weave these activities into everyday routines. With over 14,000 participants from one community contributing to the Pathway Project between 2017 and 2023, countless stories demonstrate the many ways to make each Landmark a vibrant part of daily life. We've also included inspiring quotes and reflections from those who have experienced the impact of the Landmarks firsthand. Join us and imagine a healthy, vibrant world as we travel the Pathway together…

© andreusk / Adobe Stock

Landmarks for the Early Years (Birth to 3 Years)

Birth to 3 Years: Characteristics of This Age Group

In the first few months of life, babies learn more rapidly than at any other time. Everything is fresh and everything is new. As their eyes begin to focus shortly after birth, they begin to respond to faces, and they eagerly reach out to grasp everything around them. The natural world is filled with a rich tapestry of sights, sounds, smells, and textures that invite engagement. The desire to be active in the world spurs great physical advances—rolling over, sitting up, crawling, and finally, the ability to walk!

Babies' brains are wired to notice and categorize details through their senses—that is how they make "sense" of the world around them. Allison Gopnik calls this "lantern consciousness," a state of being "lit up" and tuned into everything that is occurring around them. Repeated experiences outdoors create a stimulating environment that fosters exploration and cognitive development and lays the foundation for a deep and lasting connection to the natural world.

There is an enormous leap in the ability to explore when babies become mobile—whether crawling or walking. For this reason, the following Landmark activities for very young children are divided into "Infant" (not mobile) and "Toddler/Preschooler" (walking or crawling on their own). Both groups feel a great sense of security from the presence of their caregivers, and will quickly mimic their responses to things around them—whether that's joyful and eager or fearful and withdrawn. For this reason, it's important to bring a positive attitude of wonder and curiosity to the time you spend outdoors together.

Preschoolers are captivated by stories, music, rhythm, and rhyme, soaking in the sounds of speech as they start to express themselves in words. Songs and rhymes about animals are especially popular, helping to nurture early connections with the natural world and nonhuman friends. Time spent with a young child is a precious opportunity to slow down, unwind, laugh, and explore the wonders of the world together.

Credit: Sarah Langer

BIRTH–3 YRS

Landmark 1

Explore outdoors together at least twice each week and more if you can!

Why?

Visiting natural areas regularly benefits everyone—adults and children alike. Regardless of the weather or the season, time outside makes everyone healthier and opens our eyes to nature's wonders. Helping a child feel at home and comfortable in a natural setting, even from their earliest hours, prepares a lifelong foundation for caring, connecting, and belonging. Parents and caregivers are important role models to encourage discovery and to show respect for the living systems that sustain us. Help your child be amazed by the natural world by being amazed yourself. Give your gift of time and go outside with your child—observe and explore and wonder together.

Climate Change Connections

Whenever you engage in activities that don't require outside energy sources (walking instead of driving a car, etc.), you help your child learn to live a more climate-friendly lifestyle. All of the following activities are climate-friendly. When exploring outdoors, look for nearby places within walking distance, whenever possible.

How?

Infants

- **On a Blanket:** Lie on a blanket and watch the clouds; lie in the grass and look into the branches of a tree.
- **Tummy Time:** Encourage time on the tummy and allow your infant to sit or crawl directly on grass or a soft bed of moss.
- **Stroller Explorer:** Be a stroller explorer and explore your neighborhood and local park. Don't forget to stop and take your little one out of the stroller to experience what you encounter along the way.
- **Sunlight and Shadows:** Place your infant in a pool of sunlight and watch the dust angels float by. Show your infant the play of shadows and light on the ground. Create shadows together with your hands and arms or with a natural object, helping them notice the shifting patterns of light and dark.
- **Natural Touches:** Let your infant touch and feel various natural objects: a feather, a clump of moss, a leaf, a smooth stone, a stick, a shell.
- **Natural Sounds:** Sit quietly with your infant and together listen to the sounds of nature—the rustling of leaves, chirping of birds, or the gentle hum of insects. Can you

imitate some of these sounds in a gentle song or hum?

● **Smell the Flowers:** Hold your infant close to different flowers, herbs, or leaves and let them experience various natural scents.

● **Water Play:** On a warm day, bring your infant near a stream, puddle, or shallow basin of water. Let them touch the water with their hands or feet, feeling the coolness and movement.

● **Nature Bath:** On a warm day, bring a small portable tub or pool outside and give your infant a gentle bath in nature.

● **Bug Watching:** Find a safe spot where insects like ants, beetles, or butterflies are active. Hold your infant close and let them watch the small creatures move.

Toddlers/Preschoolers

● **Curiosity and Wonder:** Tap into your child's natural curiosity and sense of wonder; have fun and ask questions such as how do you think this got here, where is this from, who was visiting and what do you think happened? To encourage wonder, use the language of wonder. Phrases like "Wow! Is that ever amazing!" and "Did you see that?" encourage your child to ask their own questions and show interest in where their curiosity takes them.

● **Familiarity:** Visit the same places frequently, to build familiarity and comfort.

● **Clothing:** Dress for all weather conditions; raincoats and boots can make rainy days fun!

● **Play and Imagination:** Provide time, space, and materials for imaginative play and engagement. Play make-believe,

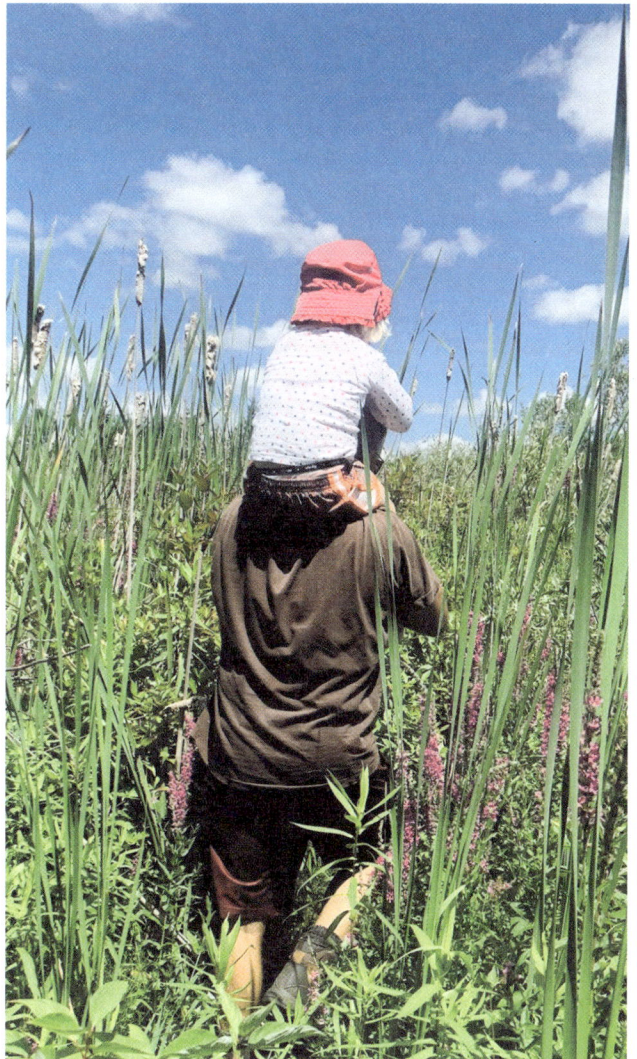

> We must teach our children to smell the earth, to taste the rain, to touch the wind, to see things grow, to hear the sun rise and night fall—to care.
> — JOHN CLEAL (artist)

Credit: Matt Warren

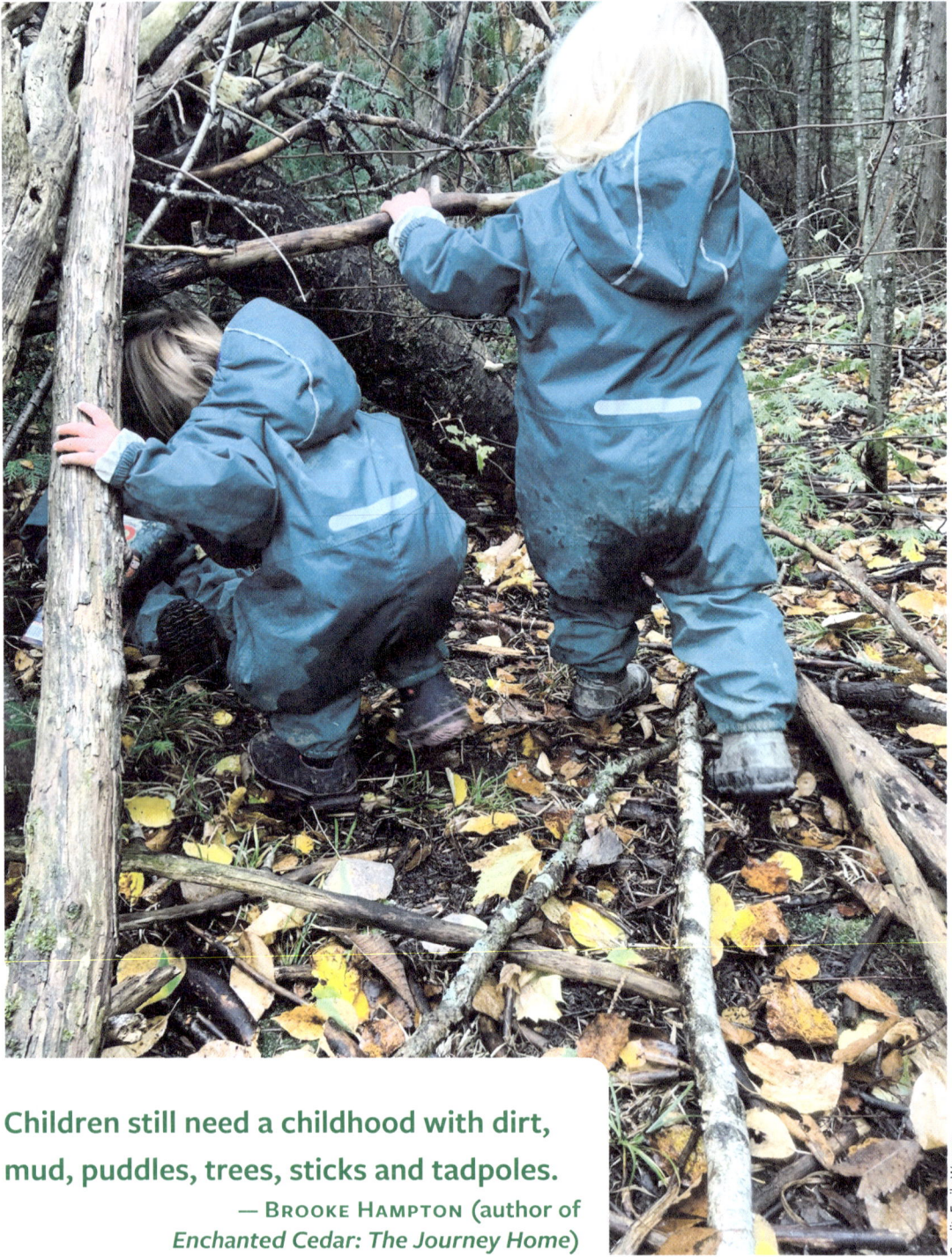

Children still need a childhood with dirt, mud, puddles, trees, sticks and tadpoles.
— BROOKE HAMPTON (author of *Enchanted Cedar: The Journey Home*)

Credit: Emily Warren

Credit: Nancy Doherty

The rain spattered our play, and the wind was wild! This weather didn't interfere with our fun though, and our climbing feet were eager, and our minds were curious.
— EMILY WARREN (early childhood educator)

explore, and imagine. Dig in soil, wade in water, pick up sticks, roll in the grass, squeeze and play with mud. Climb, jump, hop, roll, laugh! Enjoy "puddle duck days" in the rain together. Jump in leaves, play hide-and-seek. Experience rain and snow, smell flowers, splash in mud, feel tree bark, catch insects, dance together!

● **Experience as Teacher:** Focus on experiences rather than teaching—let your child take the lead. Explore together but go at the child's pace. Allow them to show you what interests them and delight in their discoveries.

● **Nature Treasures:** Create a nature treasure box or a wonder box. Decorate your container and use this to store your precious finds—a feather, a colorful stone, a shell, a four-leaf clover.

● **The Underworld:** Turn over rocks and logs to see what lives underneath (put them back when you're done). Follow a beetle, an ant, a worm—where do they go?

● **Mud Play:** Make mud pies and decorate with flowers, grasses, and leaves.

BIRTH–3 YRS
Landmark 2

Have positive experiences with animals at least twice each month.

Why?

Caring for animals helps young children develop empathy, care, and kindness. Many of us remember the joy of taking care of an animal when we were young. These experiences teach children to understand and appreciate the needs of other living things, whether it's a pet, a bird, or even a small insect.

By interacting gently with animals, children learn to be compassionate and respectful. These early experiences can prevent fears and cultivate love for the natural world. Just like a muscle, empathy grows stronger with practice, and taking care of a living creature is one simple way to promote caring and compassion. Starting this habit early lays the foundation for a lifelong respect for animals and the environment.

Climate Change Connections

Encouraging children to be mindful of the needs of other living things is an important step in making life choices that build a healthier, more sustainable world for all of life. This can start from children's earliest years, building on their natural fascination for animals.

How?

Infants

- **Animal Interactions:** Show children how to interact with friendly dogs, cats, and other pets. Gently pet them together. If you own a dog, walk them together.

Credit: Emily Warren

Credit: Matthew Walmsley

Credit: Emily Warren

- **Birdsong:** Point out and listen to birdsong when you walk together outdoors. Sing the song together when you hear a bird using a simple and memorable tune.
- **Sounds of Nature:** Listen for natural sounds in your neighborhood. Try imitating some of them together. Can you make the sound of a squirrel, the trill of a toad, or mimic the patter of rain?
- **Animal Watching:** Watch local squirrels when you're outdoors and say hello! See how they leap from branch to branch and how quickly they scurry up trees. Follow them with your eyes. Can you find their drey—or their leafy home? What other animals do you notice? Can you find a frog or even a salamander?

> If you talk to the animals, they will talk with you and you will know each other. If you do not talk to them, you will not know them and what you do not know, you will fear. What one fears, one destroys.
>
> — Chief Dan George,
> Tsleil-Waututh Nation

Credit: Emily Warren

You can learn so much from animals. They have this wonderful quality of being in the moment, and they help you spend time there.
—EMMYLOU HARRIS (singer-songwriter)

Toddlers/Preschoolers

● **Watch Beetles, Ants, and Other Insects:** Look for insects and if you find some, just sit and watch. What do you think they are doing? If you find ants, try to follow them. Where do they go? They follow one another using their antennae to pick up the scent trail left by the ants ahead. Rub your finger across the trail and watch how ants stop, and how they are able to pick up the trail once again.

● **Worm Adventures:** Dig in the soil to look for worms; hold them in your hand. Watch how they move. Try "worm fiddling." Find a stick as long as your arm. Place this in the ground after a rainy day. Rub another stick up and down against this stick repeatedly and sometimes worms will rise! Find out more about the art of worm fiddling or worm grunting.

- **Care for a Pet:** Volunteer to help to take care of a family or classroom pet.
- **Farm Visit:** Find a farm nearby that invites families to interact with animals. Watch chickens, pigs, and cows. Pet a goat or a rabbit if you can.
- **Pinecone Feeder:** Make simple pinecone feeders for birds and squirrels. Collect dried pinecones and smear them with vegetable or animal shortening and roll in birdseed. Hang up in a nearby tree. Who comes to visit?
- **Baby Animals:** Watch baby animals in nature and talk about what they're doing. Depending on where you are, it is fun to see goslings and ducklings grow so quickly. Notice how well they are able to imitate their parents.
- **Nighttime Nature Walk:** Take your toddler on a short nighttime walk to listen for nocturnal animals like crickets, owls, or frogs. Bring a flashlight to safely explore the surroundings. Sit comfortably somewhere dark and turn off your flashlight. Can you see the stars?

> **As we ventured through a new path, we were absolutely delighted to find the tiniest toad we had ever seen!**
> —MATT WARREN (parent)

> **We practiced being kind to our pet rabbit. We talked about how to pet nicely and help to feed, and not to mix food and water together.**
> —MATTHEW WALMSLEY (parent)

- **Storytime with Animal Puppets:** Use animal puppets to bring animal-themed books to life. You might be able to borrow puppets from a Family Resource Center. Act out simple stories or make up your own adventures, allowing your toddler to interact with the puppets and learn about different animals in a playful way.
- **Read Books Together:** Select colorful books featuring stories and pictures about animals and read these in a special place, outdoors if you can.

BIRTH–3 YEARS

Landmark 3

Exercise the senses every day.

Why?

Young children are sensory beings. Giving young children as many opportunities as possible to see, hear, touch, feel, and smell the environment helps them to be truly connected to the world around them. However, these days even our youngest children experience much of their lives through a glowing two-dimensional screen (computer, TV, iPad, smartphone, etc.). Only two of their senses are activated, that of sight and sound. Our children are experiencing what some have called "sensory anesthesia," the dulling of those senses not used.

Young children are more fully immersed in their environment when all of their senses are primed and activated. The feeling of wind against a child's cheek, the smell of wildflowers, or the sound of birdsong offer rich opportunities to engage the senses and forge a deeper relationship with the natural world. Look for daily opportunities to exercise a child's senses through temperature (weather and seasons), music, rhythms and rhymes, natural sounds, feeling textures, noticing smells in the sunshine or after a rain, balancing to walk across a log; feeling and tasting the snow—the possibilities are endless!

Climate Change Connections

If children spend more time in the virtual world than in the real one, they are deprived of experiencing what it means to be fully alive. Exercising the senses every day opens the doors to appreciating beauty, feeling wonder, and finding connection to the miracles that surround us. This is fundamental to developing stewardship skills and making wise choices in later life.

Credit: Stephanie Springett

Credit: Matthew Walmsley

One touch of nature makes the whole world kin.
— WILLIAM SHAKESPEARE, Troilus and Cressida

How?

Infants

● **Making Natural Music:** Provide sticks or dowels to bang on various surfaces to make sounds. Create a musical composition with your child as you clack together stones, rub your feet through dried leaves, create shakers out of containers with sand, pebbles, and small stones. Experiment with different natural materials to create unique sounds.

● **Sing Together:** Sing calming songs and action songs together. Use simple melodies, rhymes, and rhythm. Try to find rhymes and songs with a nature theme and repeat them frequently.

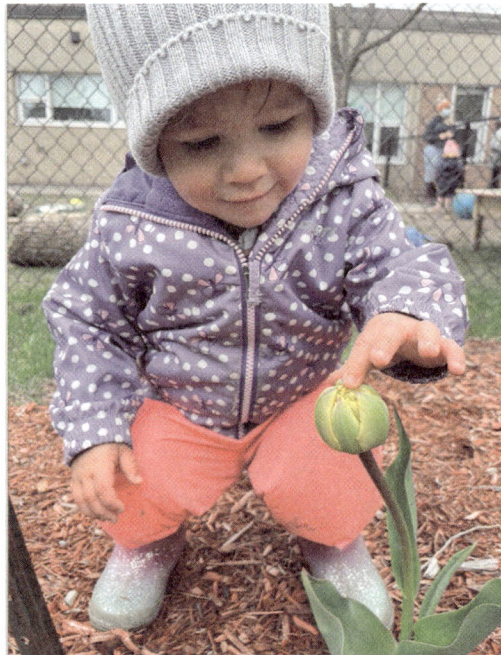

Credit: Grace Beaumont

As a child, one has that magical capacity to move among the many eras of the earth; to see the land as an animal does; to experience the sky from the perspective of a flower or a bee; to feel the earth quiver and breathe beneath us; to know a hundred different smells of mud and listen unselfconsciously to the sighing of the trees.

—VALERIE ANDREWS (author of *A Passion for This Earth*)

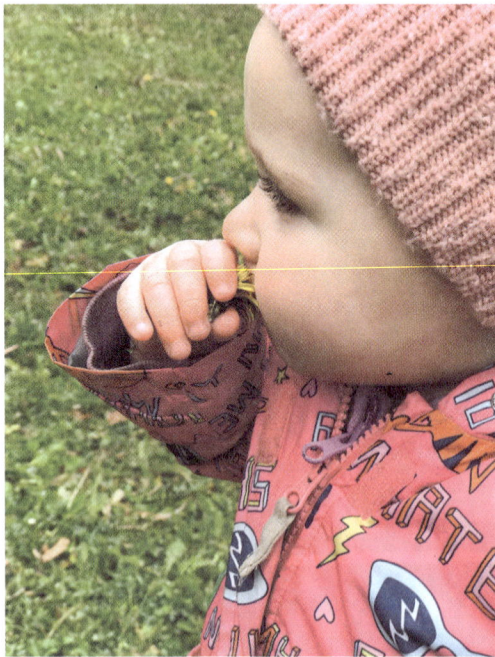

Credit: Stephanie Springett

- **Natural Textures:** Wiggle toes in the grass, in the sand, in puddles, in the soil. Allow your infant to feel different textures. Try smooth stones, warm sand, shells, sticks with rough and smooth bark, moss, leaves of different sizes and thickness, tickle with grass, squeeze mud.
- **Nature Smells:** Crush fragrant herbs in your hands and have your infant smell these. Basil, thyme, rosemary, lovage, lemon balm, anise—all have wonderful scents that are released when crushed. Smell ingredients while preparing for a meal.
- **Cloud Watching:** Let your infant lie in the grass and watch the clouds.
- **A Touch of Water:** Gently lower your infants' feet into some cool water of a river, lake, or stream.
- **Nature Mobile:** Dangle interesting shaped and colored leaves, sticks, and shells to hang over your infant's crib/bed.

Toddlers/Preschoolers
- **Mud Kitchen:** Play in the mud after a rain, or with sand at the beach. Create a mud kitchen using old pots, pans, spatulas, spoons, and pie plates.
- **Barefoot Walk:** In an area that is safe, walk through the coolness of the forest and the warmth of a meadow with your bare feet.
- **Scratch and Sniff:** Go on a scratch and sniff excursion. Use a sponge to wet the area between your upper lip and nose. The moistness helps you to smell better. While you walk, gently rub various natural

objects with your fingers to release the scent and have your toddler smell what you touch (so you know it is safe). You might try bark, leaves, different conifer needles (cedar, pine, hemlock, spruce—each have a different smell). Does the soil of a forest have a different smell than the soil of a meadow?

- **Nature Scents:** Fill small cups with various scents from the forest and from the garden. Close your eyes, can you identify what you've smelled?
- **Sensory Garden:** Create a sensory garden with different smells and textures (see page 96 Landmark 15).
- **Deer Ears:** Turn your ears into deer ears by cupping your hands, placing them behind your ears and pushing forward. You really can hear better this way! What natural sounds can you hear with your deer ears?
- **All-Around Watcher:** So many of us when we walk, we look at our feet. Instead, practice being an all-around watcher. Look in front, behind, and off to the sides. By moving our eyes around, we often spot things we might have missed.
- **Stone Soup:** This is based on a traditional story. A stranger walks into a town where the people are hungry. He says: "I will help you with this magic stone." And he drops a stone into a boiling pot of water. The stranger asks people to make a small contribution to the stone soup. He asks one person if they have just one onion and to drop this in the soup, another if they have a few stalks of celery, another a few old po-

> **The ability to touch, see, smell, and freely explore with our hands and bodies, brings not only joy but engages the soul and grounds our bodies.**
> —STEPHANIE SPRINGETT (early childhood educator)

tatoes, another a carrot, another a beet and so on. Soon there was delicious seasonal soup made by everyone, and they ate their fill. Can you make a seasonal soup of things from the garden or from the farmer's market—and use a smooth stone to start you off?

- **Nature Play:** Make a simple seesaw with a board over a log. Place a log sideways in the ground to balance on. Create a simple obstacle course using natural materials. Play hide-and-seek. Jump in a pile of leaves!
- **Wet Wood Walk:** Walk in the woods after a rain—feel and smell bark, moss, twigs.
- **Feely Blanket:** Place natural objects such as a pinecone, interestingly textured rocks, sticks, moss, leaves under a blanket. Have your toddler lie on their belly and slip their hands under the blanket. Can they guess what they are feeling by touch alone? Or use an old pillowcase and have toddlers reach their hands inside to see if they can guess what the objects are just using their sense of touch.
- **Snow Play:** In winter, create simple sculptures out of packing snow. Or bring basins of snow indoors in winter to feel and shape, using spoons or spatulas.

Landmarks for Age 4 to 5 Years

Age 4 to 5 Years: Characteristics of This Age Group

Children aged 4 to 5 years are in a stage of rapid development, where they are becoming more mobile and independent. Children at this age are naturally curious and eager to explore. They often ask "why" questions and enjoy hands-on activities that allow them to interact directly with their environment. They learn best through play, observation, and sensory experiences, making outdoor activities ideal for learning

Credit: Emily Warren

about the world around them. Imagination is a key part of their cognitive development. They enjoy self-directed play, stories, role-playing, and make-believe, which can be used to introduce concepts about nature and the environment in a relatable way. Their imaginative thinking allows them to empathize with animals, plants, and other elements of nature, often personifying them, which can foster a sense of connection and care.

At this age, children begin to understand the feelings and needs of others, which extends to animals and plants, and they can be encouraged to use words to express their own feelings too. They can start to appreciate the importance of caring for living things. Activities that involve nurturing, such as watering plants or feeding animals, help strengthen this emerging empathy. They continue to carefully observe the behavior of adults and older children, often mimicking their actions. This is an ideal age to observe others modelling respectful interactions with other living things.

Children at this age are beginning to assert their independence and enjoy tasks that allow them to feel capable and responsible. They love to be praised for helping others, which reinforces the idea that their actions matter. They learn best from their immediate surroundings. Regular outdoor play in diverse natural settings—like parks, gardens, and backyards—can spark their interest and deepen their connection to nature.

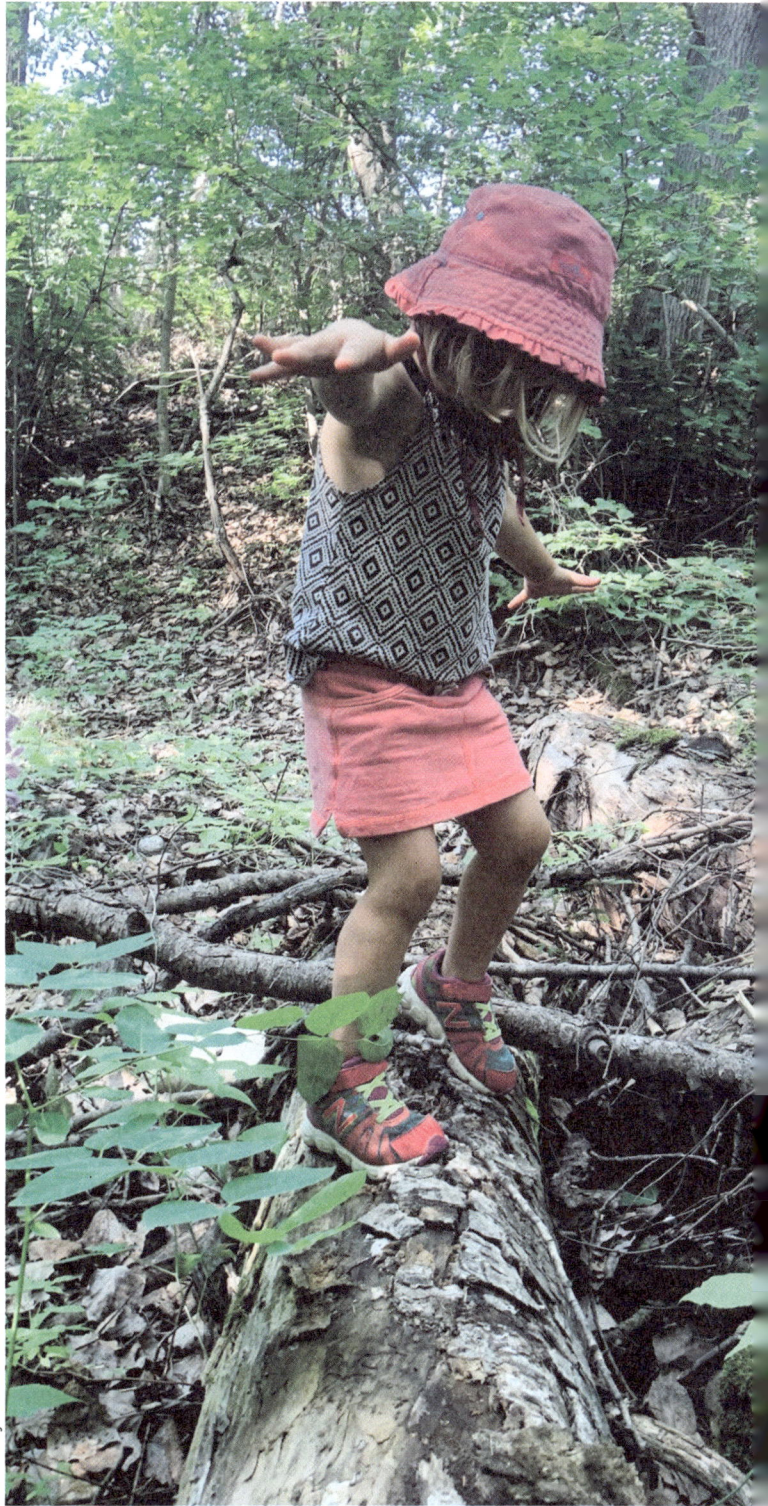

Credit: Emily Warren

AGE 4–5 YEARS

Landmark 4

Visit a favorite outdoor place each week throughout all seasons. Share what you have discovered with a supportive adult.

Why?

We were all born to connect to the world around us. And that sense of belonging to something bigger than ourselves is enhanced when we become familiar with a favorite place. When we visit often, we come to know the stories of that land and the characters that live there. For example, that heart-shaped hole is where a squirrel builds its nest and who stamps its feet and chatters at us when we get too close; that rock was left by the glaciers thousands of years ago and is filled with quartz crystals, sparkling in the sun. Every natural place is filled with stories and opportunities to explore and discover. Parents and teachers can promote a sense of awe and wonder by taking the time to honor and celebrate a child's discovery.

In the spring, at the end of the day, you should smell like dirt.
—MARGARET ATWOOD (author)

When children are dressed for the weather, they can experience the magic of their favorite places in all their transformations—rain, snow, sunshine, fog, and wind—through every season. As they return to these familiar spaces, a deep sense of security and belonging emerges, allowing the land itself to become their most powerful teacher. At the same time, children develop resiliency and become more in tune with the rhythms of the changing seasons.

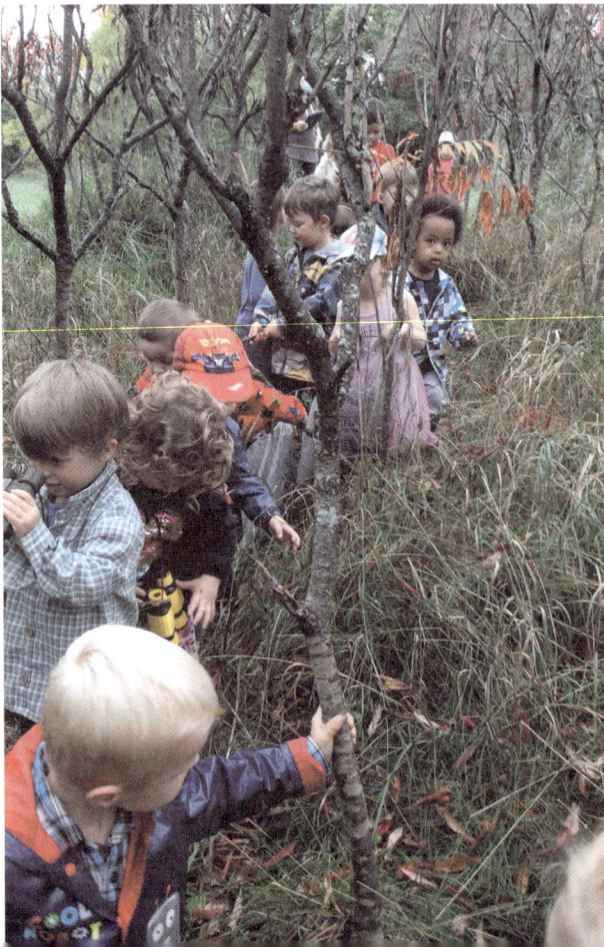

Credit: Stephanie Springett

Recipe to raise a Wild Child: "Take one child. Place outdoors in nearby green spaces. Leave for several hours at a time. Repeat daily. Sprinkle in a dash of adventure. Fold in a generous portion of exploration and discovery. Top with wonder and awe. Let rise…"

—Jacob Rodenburg, from *The Big Book of Nature Activities*

Credit: Emily Warren

Climate Change Connections

Look for special places that are close to home, or that you can visit by bicycle or bus. Try to minimize fossil fuel consumption whenever possible while exploring each Landmark. This is positive action to reduce climate change!

How?

● **Special Spot:** Visit a favorite climbing tree or balancing log throughout the year. Find a secret magic spot that is sheltered, perhaps under the boughs of a pine tree or in a bush. Go to this space each time you visit. How does it change through different kinds of weather, through different seasons? In your magic spot, sit quietly and really observe. What does this spot teach you about the world around you? Did you notice a caterpillar munching on leaves, a bird singing from a nearby tree, or different decomposers such as mushrooms and worms gradually returning the forest litter to soil?

● **Build a Fort:** Try building a sturdy shelter with found materials: sticks, boughs, leaves, and branches.

● **Hide-and-Seek:** Look for places to hide in your favorite place; play hide-and-seek.

● **Wonder Wagon:** Create a "wonder wagon" or a "wonder backpack." Fill it with a few nets, binoculars, magnifying glasses, string, crayons and paper for rubbings and drawings. Take this with you when you explore your favorite natural place.

Credit: Emily Warren

● **New Ways to Travel:** Find new ways to travel to your favorite place: try biking, try sliding on an icy day.

● **Visit a Nearby Park or Conservation Area:** Explore a trail through each of the seasons. Take your time. Gently lift rocks and logs and peer underneath. What kinds of interesting critters live there? We call these "basement windows," a portal into the magic underworld of soil. Don't forget to carefully close the window when you are finished looking.

Every day we explore our Habitat area on the school yard together. This is an ongoing daily activity. We use our wonder wagon complete with magnifying glasses, tape for nature bracelets, bungee cords and skipping ropes for tying, mud kitchen accessories, clipboards, paper, crayons to document our findings.

—Jackie Mercer (teacher)

● **Found Natural Objects:** Make a rainbow out of found natural objects. Collect objects of different colors: stones, sticks, leaves, bark. Arrange from darkest to lightest, biggest to smallest, curvy to straight. Collect as many different shaped leaves as you can. Make a nature alphabet out of found natural objects.

● **Rubbings:** Use crayons and paper to create rubbings of leaves, bark, or interestingly textured rocks.

● **Simple Nature Journal:** Keep a journal and make simple drawings of what you see each time you visit.

● **Nature Table:** Gather a few natural objects to take home to a nature table—a colorful stone, an interestingly shaped stick, or an old nest. Make sure to return these back to the land when you are finished with them.

● **Beach:** Visit the beach during the winter. Bring a winter picnic. How has the beach changed since you visited it in the summer?

© ngamaz / Adobe Stock

AGE 4–5 YEARS

Landmark 5

Help to plant or harvest a garden and/or look after an animal.

Why?

There is something magical about planting a garden, watching a seed sprout, digging in the soil, watering, tending, and then harvesting fresh produce straight from the earth and into your mouth.

Learning to care for the needs of living things by practicing gardening or looking after an animal is a wonderful way to develop empathy. Children can be encouraged to expand their understanding of well-being by thinking about what other things need to grow and thrive. Young children have a limited sense of time, so helping with short-term tasks are ideal at this stage of development. They can begin to understand that we all need similar things: food, water, a safe place to live, exercise, companionship, and kindness. Caregivers are important mentors in modelling caring relationships. Developing positive relationships with the natural world lays the foundation for an ethic of stewardship that will endure long into adulthood.

Climate Change Connections

Making locally grown food a major part of your diet makes a positive contribution to combatting climate change. Global food transport is a major source of greenhouse gases, so buying local or growing your own together is taking a positive, constructive step together in the right direction! Giving children the opportunity to help grow food develops an important, sustainable life skill.

How?

● **Plant Veggies:** Plant some simple vegetables with big seeds: beans, squash, and peas are great plants for a young child to grow.

● **Hand-Feeding Chickadees:** Try feeding chickadees with sunflower seeds in your hand; it takes patience to win their trust at first, but later they will visit eagerly! Black oil sunflower seeds seem to work best. If chickadees don't live near you, are there other kinds of small friendly birds who will take seeds from your hand?

● **Grow Indoor Flowers:** Grow an indoor pot of sunflowers, marigolds, or spring bulbs for beautiful flowers.

Credit: Emily Warren

When you teach a child to garden, you
show them how to grow their own future.

—ANONYMOUS

● **Hatch Chicken or Duck Eggs:** Often farmers will loan you an incubator and some chicken or duck eggs so that you can watch eggs hatch out and spend some time caring for the chicks before returning them to the farmer.

● **Care for Worms:** Learn about worms and how they help to decompose food and transform this into soil by setting up a classroom vermicomposter.

● **Compost Magic:** Together start a compost bin with kitchen scraps and yard waste. Learn about the composting process and how it benefits the garden.

● **Create an Eggshell Garden**: Adults—purchase a dozen fresh eggs. Use a sharp knife to cut off the top third of the eggs and then pour the egg batter into a bowl (don't forget to cook these up for breakfast!). Rinse your eggshells with water. Use a needle to poke a hole in the bottom of the shell so that your egg container can drain. With your kids, place your empty eggshell back into the egg carton and add potting soil. Press small seeds a little way into the soil and sprinkle more soil on top. Give another spray of water and place your egg carton in a sunny spot and see what happens! When your eggs sprout, you can place the whole egg with your seedling into the ground. The eggshells will eventually break down and biodegrade adding nutrients to the soil.

● **Herb Planter Box:** Plant a variety of herbs in a small planter box or pot and place in a sunny windowsill. Use the herbs in cooking to show how growing your own food can be both fun and useful.

● **Mini Green House:** Use a plastic bottle or container to create a mini greenhouse for seedlings. Observe how the greenhouse effect helps plants to grow.

● **Bee Hotel:** Build a bee hotel to provide a habitat for solitary bees. Learn about the different types of bees that live near you and their roles in pollination.

● **A Flutterby:** Witnessing the transformation of a caterpillar into a butterfly is a magical experience. Raise a caterpillar and release the butterfly! Make sure to research what food your caterpillar needs to grow and thrive.

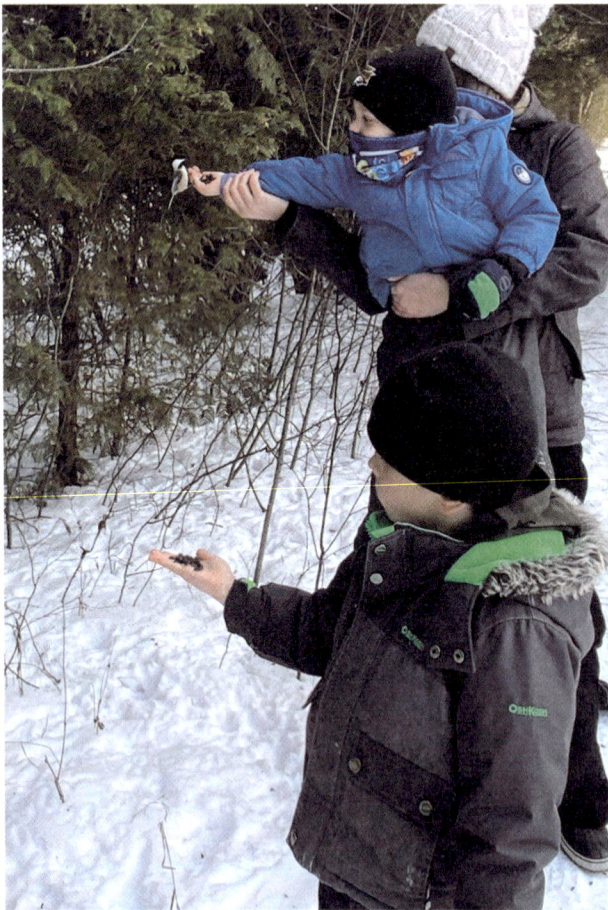

Credit: Shelley Tremblett

Watching the chicks this week has prompted lots of thinking about growing and changing. We noticed the chicks have lost their egg teeth. Some students shared how they have lost some baby teeth. We watched as the chicks cleaned their feathers with their beaks. The kindergartens talked about how they care for their own hair by brushing it. We also noticed how the chicks' wings are growing and they are now able to fly up and around!

—Indrani Talapatra (teacher)

- **Flowers:** Plant edible flowers like nasturtiums or pansies. Use them to decorate dishes and learn about the connection between gardening and cuisine.
- **Dry Garden:** Create a garden with a variety of succulents. Learn about drought-tolerant plants and their care.
- **Birdbath:** Try setting up a birdbath or toad pool in your yard during warm weather. Who visits it? Remember to clean it every week and add fresh water.
- **Feeders:** Hang up birdfeeders. Use nyjer seed to attract finches, mixed seed to attract a variety of birds, suet feeders for woodpeckers, and platform feeders so squirrels can have a snack as well. You can also create a simple birdfeeder out of an empty milk carton. Simply cut a hole halfway up, poke a short stick underneath the hole for a perch, and fill with black oil sunflower seeds.

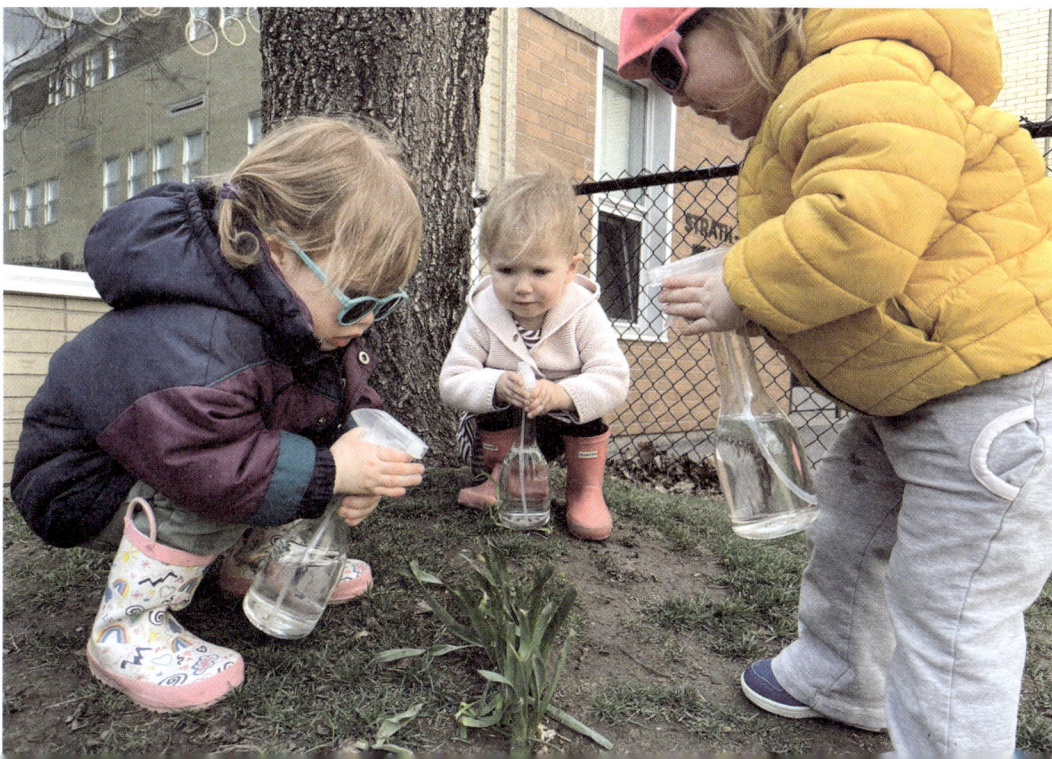

Credit: Grace Beaumont

AGES 4–5 YEARS

Landmark 6

Play in nature for a full hour at least twice a week, and more if you can.

Why?

What happens when children play in nature? So many things in the natural world lend themselves to imaginative play. A stick becomes a sailing mast or a magic wand. That moss covered tree transforms into a spaceship. The area around a bush becomes a castle. Giving children the gift of time and opportunity to play in nature shapes and strengthens their imagination. Playing together helps kids socially as they learn to negotiate, deal with conflict, take turns, problem solve, and initiate play. This in turn fosters independence and resiliency. As a parent or educator, instead of being a helicopter that is always hovering and monitoring, try being a hummingbird, where you flit in, to check on things, and then you flit away. Renowned psychologist Jean Piaget tells us that, "Play is the work of childhood."

When children become familiar and feel safe and comfortable in a place, nature provides materials to nurture the imagination, and boundless stories begin to unfold. Giving children at least an hour to be immersed in a natural setting helps them to begin to feel that this is their home too!

Climate Change Connections

All of these activities are climate-friendly, especially if they take place within walking or cycling distance of home or school.

How?

● **Mud Kitchen:** Provide pots, pans, containers, soil, and other natural materials for children to create imaginary food. Fill an empty aluminum pie shell with mud and decorate the top with flower petals, colorful pebbles, shells, or leaves to create amazing mud pies!

● **Tiny Playground:** Using small sticks, pebbles, and string, build a tiny playground for a beetle or a caterpillar. Can you make slides, swings, balance beams, and a small sandbox?

● **Big Nest Building:** Can you make a nest for humans? Gather leaves and dried grass. Mound these up and hollow out the center. For an extra challenge, use grapevine and upright sticks and weave a more permanent nest. Have an adult help you. Share stories while you are cozy and huddled in your nest.

● **Insulated Fort:** Using a variety of sticks, make an outdoor fort. Start with a ridge-pole and place sticks on either side. When you have a stick frame, pile on dried grass, evergreen boughs, and leaves to insulate your structure. Don't forget to insulate the bottom of your fort with natural materials.

- **Fairy Gardens:** Create a fairy garden or a gnome home with found materials.
- **Obstacle Course:** Make an obstacle course with tires, logs, straw bales, and whatever you can find.
- **Nature Faces:** Make nature faces using natural materials. Press clay on the side of a tree to make faces and decorate with leaves, twigs, sticks, and moss. Then, go out and look for other faces in nature. Are there faces in tree bark? in patterns of leaves or stones? in clouds in the sky?
- **Tiny Scavenger Hunt:** Can you find the following and fit all of these into one small container? Explore with your child and ask them to find something that is:
  - *Beautiful*
  - *Sparkly*
  - *Soft*
  - *Hard*
  - *Warm*
  - *Has many colors*
  - *Older than you*
  - *Younger than you*
  - *Has one color*
  - *Doesn't belong*
  - *Cold*
- **Loose Parts:** Provide a whole collection of loose parts, pinecones, smooth stones, boughs, tree slices, branches, shells. Place each of these in a basket and let children use their imagination to find different ways to engage and play with this material.

> ## Children need to be outside long enough to feel at home there.
> —**EMMA SHAW** (founder, Into the Woods Outdoor Nursery)

- **Egg Carton Collection:** Over time collect your empty egg carton boxes. When you have enough for each child, go and explore. Try to fill each compartment of the egg container with something special from nature: a sparkling rock, an unusual looking leaf, a bit of moss or lichen, a beautiful flower (only harvest small bits, and only if there are many more).
- **Twig Boat:** Gather a dozen or so twigs that are about 6 to 8 inches long. Tie them together in a bundle so you create a raft. Place a small dab of clay or plasticine in the middle. Stick one straight twig in the center of this so it acts like a mast. Bend a

Credit: Indrani Talapatra

Encourage your child to have muddy, grassy or sandy feet by the end of each day. That's the childhood they deserve.

—PENNY WHITEHOUSE
(founder, Mother Natured)

large leaf and poke this through the mast, one hole through the top and the other through the bottom to act as a sail. Find a river, stream, or pond (even a bathtub will do) and begin your voyage!

● **Animal Footprint Cast:** Take some plaster of paris (available in any hardware store) with you on a hike, a spoon, a container of water, and an empty container (used yogurt containers work well). Look for a clear animal track in the mud. Mix just a bit of plaster and water together until it is smooth and the consistency of porridge. Pour directly into the track. Wait about 40 minutes. When you return, gently ease the track out and rinse with water. You'll have a nice cast of a real animal footprint!

Credit: Nancy Doherty

The puddles of our yard bring us immeasurable joy! These friends see a rainy day as an opportunity and an invitation to play as freely as their hearts can handle. Jumping harder and bigger in the puddles than the last day, running and sliding quicker and with even more exuberance. What a delight to watch these friends enjoy themselves so thoroughly!

— Emily Warren (early childhood educator)

Credit: Stephanie Springett

AGES 4–5 YEARS

Landmark 7

Share a nature-based picture book, song, poem, or game each week.

Why?

Stories, pictures, songs, and games help children love and understand the natural world. Our brains are hardwired to listen to stories; after all, for thousands of years before the written word, this is how knowledge was transferred from one generation to the next. Stories are a universal form of communication that resonates deeply with us. We all love stories—they're a wonderful learning tool that stimulates the imagination, conveys complex ideas, and builds loving relationships. Songs, dance, rhymes, and games further enrich these connections, bringing joy and excitement as we celebrate the amazing world we share. Young children have a natural affinity for rhythm and rhyme, and take great delight in songs and rhymes that have actions associated with them. Sharing nature-based materials reinforces positive associations with the living world, and stimulates the imagination.

Climate Change Connections

It is an awesome gift for children to learn the many ways we can entertain and enjoy ourselves without needing to use fossil fuels. Books, songs, poems, and nature games can be enjoyed by anyone, anywhere—even when the electricity is off, or there's no gas in the car!

How?

● **Animal Picture Books:** Look for picture books with stories about a variety of animals (especially those animals that live nearby). Ask your local librarian for suggestions for great picture books that celebrate nature or demonstrate stewardship; there are so many wonderful books to share with young children! Here are a few of our favorites: *Outside*, *You Notice*, by Erin Alladin; *Turtle Walk*, by Matt Phelan; *The Little Hummingbird*, by Michael Yahgulanaas; *The Mitten*, by Jan Brett.

● **Songs and Rhymes:** Learn songs and rhymes that celebrate animals, trees, and the changing seasons. Here are a few easy rhymes that you can do with actions together:

Credit: Candace Passey

- *Here is the earth, here is the sky, here are my friends, and here am I!*
- *Rain is falling down, splash (×2), Pitter patter, pitter patter, Rain is falling down, splash! Sun is peeking out, peek (×2), Peeking here, peeking there, Sun is peeking out, peek!*
- *Here is a nest for a bluebird, here is a hive for a bee, here is a hole for a bunny, and here is a home for me!*

- **Nature Story Books:** Nature-based story books are great to read together indoors as well as outdoors in a favorite place at all times of the year.
- **Follow a Track:** After a fresh snow, one person makes a track for others to follow;

Passion does not arrive on videotape or CD. Passion is personal. Passion is lifted from the earth itself by the muddy hands of the young; it travels along grass-stained sleeves to the heart.

—RICHARD LOUV,
in *Last Child in the Woods*

Credit: Danielle Blondin

Playing the Pumpkin Game

Tracks in the snow, tracks in the snow; who made the tracks, where do they go?
—Indrani Talapatra (teacher), from the book by Wong Herbert Yee

what story do the tracks tell? Can you find fresh tracks made by animals? Follow the track. Which way was it going? What was it doing? What story has the animal left behind?

Teaching kids to count is fine, but teaching them what counts is best.
—Bob Talbert (journalist)

● **Events:** Participate in music and story-telling events at the public library, at children's centers, or invite a storyteller to visit your class.

● **The Unfolding Story:** Start with a piece of paper. Agree with everyone that you'll create a story about a local animal. Have everyone draw a picture of this animal doing something (running away from danger, finding food, hiding). Gather the pictures and combine them into a booklet. Can you make a story of the children's combined pictures? When you have a story—share this with parents or another class.

● **Create a Felt Storyboard**. Cut out felt pieces of different local animals, insects, birds, and plants. Create a colorful backdrop of a forest or meadow. Have children create their own stories using these pieces.

● **Create a Storyland.** Use a piece of plywood 2' × 4' and trim the edges with 1" × 2" wood. Fill this area with sand, creating hills and valleys. Gather cedar or other small evergreen tips and place these into the sand all around to create a forest. Use blue cloth and place along the valley to create a river. Find small cloth or clay animals along with shells, pinecones, smooth pebbles (any natural materials) and have children

Chef Cindy created a cool hide-and-seek game for us with some "pumpkin" rocks. We had to find all 15 rocks and match them to the picture. Then we could hide them again for the next class.

—Danielle Blondin
(early childhood educator)

Credit: Heather Snowball

Playing What Doesn't Belong?

use their imaginations to create nature stories to share.

● **Create a Story Trail:** Laminate each page of your favorite nature storybook—one that has colorful pictures and isn't too long. Or, use one that only has pictures. Pin each page along a trail. Escort your children along the trail—stopping at each spot so that you can sit and follow the unfolding story in an inspiring natural setting.

● **Find a Special Story Spot in Nature:** Perhaps your special story spot is a tree with overhanging boughs that offers a bit of a refuge. Or maybe you can find a meadow with wildflowers. Or a wide branch that invites you to sit under an old oak. Try going to your special story spot over and over again to read a nature storybook or to tell a story to your friends. The more you visit your story spot the more special it will become.

Today we played a fun activity which connected our recent discussions about the five senses. Students had to travel along a short nature path looking carefully for items along the trail that didn't belong in nature. I stood at the end of the trail, and students whispered the number of items they counted to me. Everyone was very determined to spot all of the items and walked the trail over and over until they saw them all!

—Heather Snowball (teacher)

AGES 4–5 YEARS
Landmark 8

Create at least one nature art project each week.

Why?

We all have a deep desire to express ourselves, to make sense of our complex feelings and the world around us. Children, too, need this outlet for expression. Nature provides a perfect canvas for creativity, offering a rich variety of colors, textures, shapes, and patterns. Using natural materials like leaves, stones, twigs, and flowers to create art offers a hands-on, immersive experience that also fosters empathy for living things. Natural art projects can be highly collaborative, encouraging teamwork and cooperation among children. These activities promote problem-solving and critical thinking, enhancing cognitive skills while delivering a joyful and engaging experience.

Climate Change Connections

Nature-based art, created with materials found nearby, is one of the most climate-friendly forms of expression imaginable! When children learn to see the beauty in nearby nature, and shape it to express their emotions and identities, they learn that they can also shape their world in positive ways. This is all part of the process of building self-esteem, empowerment and hope.

How?

● **Magic Wands:** Paint pieces of tree branches, wrap with ribbons and leaves. Use your wand to create magic worlds in nature.

● **Frame It:** Make a frame by glueing popsicle sticks into a square. Place your frame over something eye-catching in nature—a colorful flower, an interesting rock, a hole in the ground. Stretch rope between trees in an area that has a lovely vista. Clip your frames onto the rope using clothespins. Peer through the frame to see a lovely view. Go on a gallery tour and visit all the frames.

● **Leaf Creations:** Arrange different sizes and shapes of leaves onto paper so they resemble an insect or an animal. Glue these onto watercolor paper and display!

● **Natural Paintbrushes and Ink:** Gather sticks that are about 6 inches long and ½ inch wide. Using scissors, gently harvest a clump of soft grass, a bundle of pine needles, a sprig of cedar, the tip of a mullein leaf, the ends of dandelion leaves—really anything that is soft and pliable. Using woolen yarn, thread, or wire, tie the tips of your paintbrush to the shaft of your sticks. Now all you need is some natural ink. Gather ripe berries such as raspberries, strawberries, and blackberries. Frozen

> **Those who contemplate the beauty of the earth find reserves of strength that will endure as long as life lasts.**
> —RACHEL CARSON (author)

berries work as well. Crush your berries until you have a fine pulp. Add a teaspoon of white vinegar and a dash of salt per pint of crushed berries. Simmer on a stove on low for about 1 hour. Use your brushes and ink to make beautiful creations.

● **Nature Sculptures:** Andy Goldsworthy is a remarkable artist who uses natural materials to create inspiring art. He might pin up colored leaves in the fall using thorns and allow the light to shine through. Or he might create spherical objects using intertwining branches. Goldsworthy uses material he finds in nature, and he makes art with his hands only—he doesn't use tools of any kind. He then takes a picture and allows the natural world to reclaim his creation. Use Goldsworthy's photos which you can find online to inspire your students or children to create beautiful natural sculptures. Take a picture of each and create a nature art exhibit.

● **Hammered Prints:** Carefully collect a variety of natural objects that have moistness in them. Examples might include dandelion flowers and leaves, clover, violets, and oak leaves. Place a white cloth or watercolor paper on top of a hard wooden surface. Artfully arrange a variety of flower heads and plants on the cloth/paper. Cover this with two or three layers of paper towel. Now, methodically and carefully, hammer

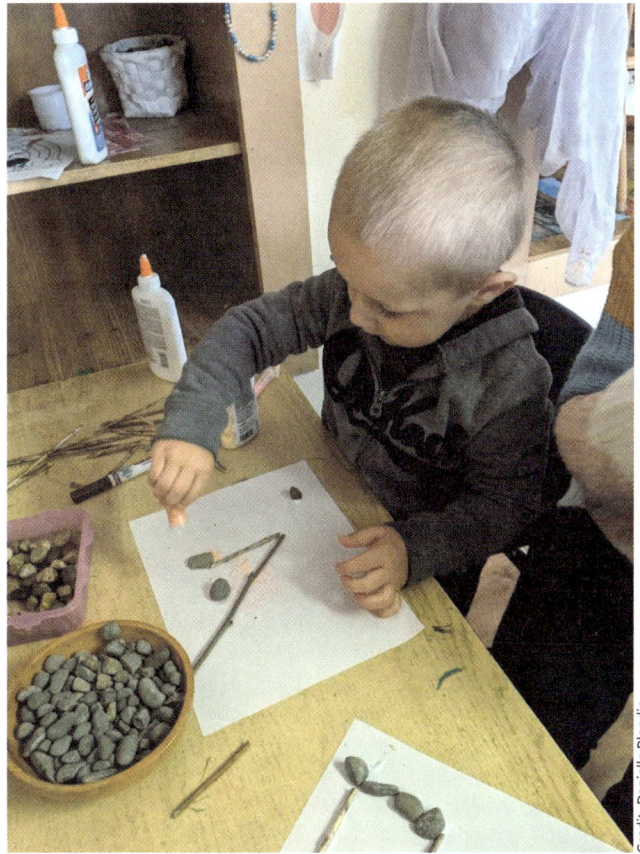
Credit: Danielle Blondin

on top of the paper towel, trying to make sure you hit each part of the plant hidden below. Careful that you don't smash your fingers! When you are done hammering, gently peel away the paper towel and the plant. You should have a lovely imprint of your plant embedded in the watercolor paper or the cloth. Admire your creation and savor the beauty of the natural world!

● **Clothespin Butterflies:** Use water color paper and decorate colorful butterfly wings. Glue these onto a clothespin (as the body) and pin these to trees or flowers outside.

● **Fall Bouquet:** Collect colorful fall flowers and make a bouquet, using a half of a small squash as a vase.

Children grow up hearing how broken the environment is, how broken beyond repair. Plant strawberries together, make wild medicines, paint the sunrise. Show them proof that for every act of destruction, they can sow a seed, however small, of beauty.

—NICOLETTE SOWDER
(creator of Wilder Child)

● **Paint Rocks:** Collect smooth rocks. Wash the rocks carefully to remove grit. Let dry and paint over with white acrylic paint. This will make your colors more vibrant. When dry, use paint to create ladybugs, animals, or whatever catches your fancy. Try making segments of a snake and joining all your rocks together. Finish by creating a snake head and tail.

● **Ice Lanterns:** When the temperature is below freezing, you can create beautiful ice lights. You'll need round balloons (any size), a large bowl, water, and tea light candles. Attach the balloon to a water faucet and fill to desired size. Tie off the balloon. Put the balloon in a bowl and place it either outside or in a freezer for six hours or until the outside of the balloon is frozen but there is still water inside. Carefully cut and peel away the balloon. Working over

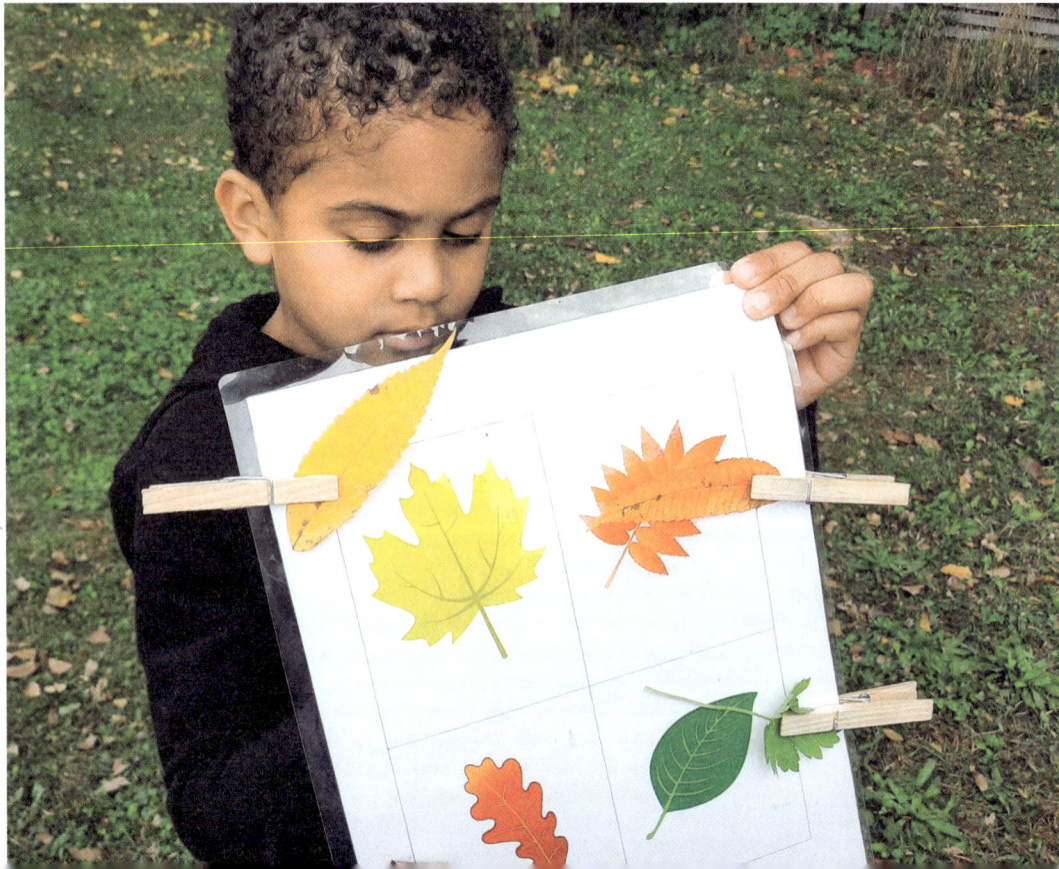

Credit: Lisa Gutoskie

a sink or outside, drain the water away by making a hole in the weakest part of the base using a kitchen knife. The hole will have to be large enough for inserting the tea candle. Refreeze the globe until completely solid. Slip in your candle and watch your ice lantern provide a warm glow in the dark of a winter's night. Make a series of these and place the ice lanterns along a walkway to light your way.

● **Frozen Mandalas:** Take sprigs of your favorite winter plants (suggestions: dogwood, rowan tree berries, wintergreen, holly, evergreens such as spruce, balsam fir, hemlock). Use a flat and shallow container (old tuna tins, pie plates). Fill with fresh water. Place your plant inside. Place a twig in the upper middle and allow this to freeze overnight. The next day, gently ease your decoration from the container. Remove the twig. Place a ribbon or twine in the hole and create a loop. This will create a handy hanger. Hang on the nearest tree. Watch as the winter light filters through the beautiful textures of ice and plant material.

● **Nature Medallions:** Collect a variety of beautiful outdoor materials (pebbles, small leaves, twigs, seeds, small flowers) with the children. Give each child a piece of modelling clay to shape into a flat medallion. They can press their nature treasures into the medallion to make beautiful colors and patterns. Medallions can be left as table ornaments, or strung onto a colorful ribbon to wear. (Remember to make a hole in the medallion for the ribbon.)

● **Snow Painting:** Use spray bottles with water and food coloring to spray beautiful patterns in the snow during winter months.

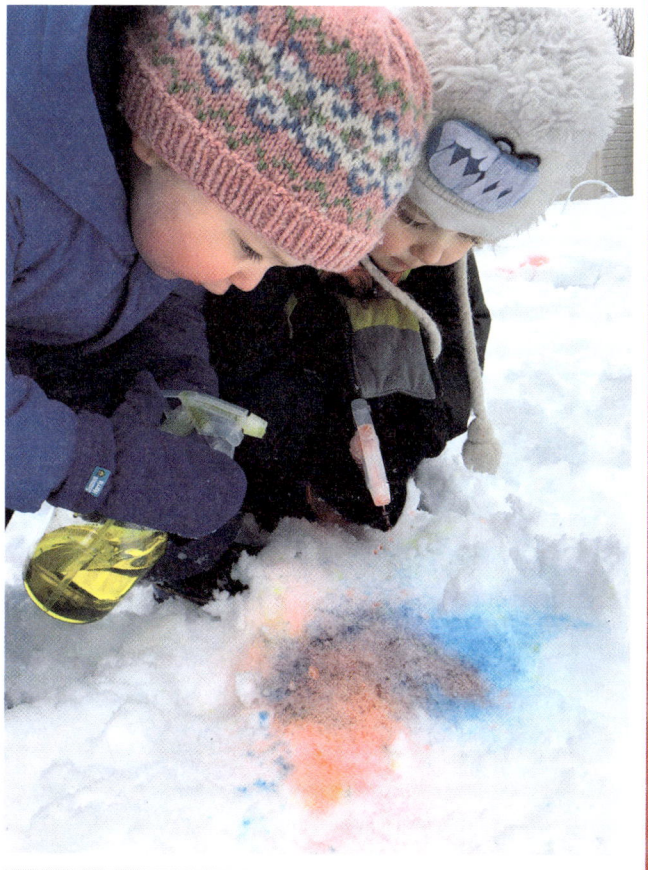

Credit: Lisa Miles

We searched the schoolyard for pieces of nature that had already fallen to the ground. We then used these items to create nature art. Once everyone was finished, we opened up our "nature art gallery" and walked around to enjoy each other's creations.
—HEATHER SNOWBALL (teacher)

Landmarks for Age 6 to 7 Years

Age 6 to 7 Years:
Characteristics of This Age Group

Children aged 6 to 7 are at a fascinating stage where their cognitive abilities, social awareness, and sense of responsibility are rapidly maturing. This age marks a transition from the early childhood years to a more structured and independent way of interacting with the world. At this age, children begin to think more concretely. They're able to follow step-by-step instructions and understand basic concepts about how things work. They enjoy solving problems and figuring out cause-and-effect relationships. This makes them more capable of understanding how their actions—like planting a tree or gardening—can have a direct impact on the environment. They

Credit: Heather Snowball

also start asking deeper questions about the world around them. "Why do trees lose their leaves?" or "What happens to animals in winter?"

Children in this age group are natural explorers, eager to investigate the world beyond their immediate surroundings. They love to explore new environments—whether it's a pond, a forest, a desert, or even a small patch of garden—and are excited by discoveries like finding a frog, watching a bird build a nest, or observing a caterpillar turn into a butterfly.

Young children also respond more to micro-environments or small areas, instead of large vistas. One of us, after a long hike with our children, finally reached the top of a towering hill. While the adults were marveling at the way the sunlight sparkled on the water of distant lakes, the kids were hunched over, studying a line of ants as they marched on by. Because they are close to the ground, younger children are more in tune with what is going on directly around them and right beneath their feet.

And of course, children at this age love to play. They are especially receptive to the organic forms of nature and love to build forts, play hide-and-seek, and make up games. This is an evocative time of imagi-

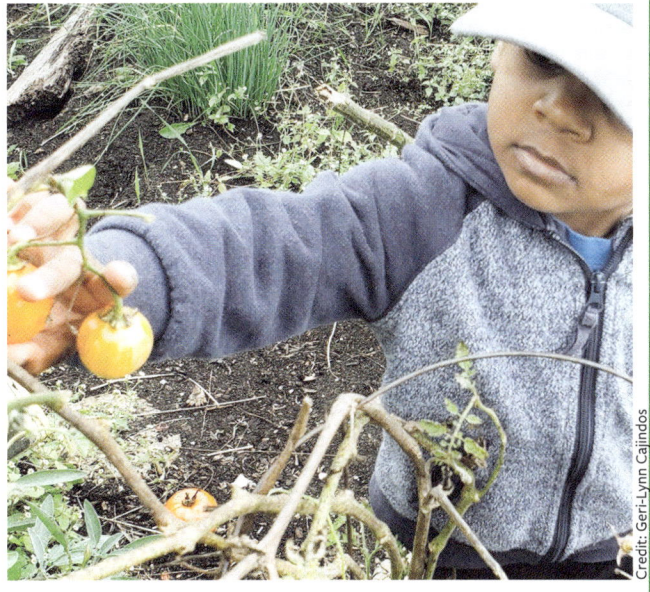

Credit: Geri-Lynn Cajindos

nation and wonder for the newness of the world around them.

Young children have an emerging social conscience that makes them more open to understanding the importance of caring for the environment. They might feel sad if they see an animal that has been hurt or when they learn that a nearby field will be dug up to make way for a housing development. They also begin to understand community roles and enjoy feeling like they're contributing members. They want to help and are motivated by the idea that their actions matter to the group, whether it's their family, class, or neighborhood.

AGES 6–7 YEARS
Landmark 9

Choose an outdoor place in nature that is special to you. Visit at least twice every month; try to visit through a whole year.

Why?

We yearn to belong to our friends and our family. But we also yearn to feel connected to a place. This helps us become anchored to the world, to belong to something bigger than ourselves, and in belonging, we feel a little less alone.

This is an important age for beginning to develop a sense of place—an outdoor space that is familiar and special. A sense of place contributes to a child's developing understanding of connection—to a family, a school, as well as a place. Children can be encouraged to deepen their relationships with special places, explore more thoroughly and fine-tune their senses: looking carefully, waiting quietly, touching gently, listening intently. Give children plenty of room for unstructured outdoor play. Adults can be supportive and share a sense of wonder, but let children decide how their play will unfold. This is a good time to gently help children overcome fears they may have of spiders, worms, snakes, or darkness, through familiarity and understanding. Giving children the chance to deeply connect with a place lays the foundation for a lifelong commitment to caring for the Earth. In the end, we will only protect and nurture what we love.

Climate Change Connections

This is one of several Landmarks that encourage developing a strong relationship with a nearby natural area—each with an extending timeframe, and more deeply observant activities as children grow older. Respectful relationships can then lead to building a sense of reciprocity—how can we give back to the places that give so much to us? Indigenous scholars teach that these are necessary steps to develop a sense of responsibility for the Earth's well-being. Sitting quietly in a place without any techno-gadgets is a skill many children have lost. It can be very calming to learn how to be still and quietly listen and observe. This kind of calm in our hectic modern world can be rejuvenating for children, and foster a deep sense of kinship.

How?

● **My Personal Spot:** Choose a spot that is special to you nearby your home. Make sure you visit this in all kinds of weather and in all seasons. Give your spot a name and name the things you see around you. Over there might be "grandfather rock," or that cedar might be "sweet tree" because of the scent of the needles. When you give your own names to beings on the land, you help to transform the natural world from a series of objects into something that is vital and alive.

● **Squirrel's Nest:** Study a squirrel's nest (also called a drey). How is it made? What materials did the squirrels use? Can you build a fort based on how a drey is made and make it large enough for a human? Use branches, leaves, and whatever natural material you can find. Is it cozy and warm? People who make survival shelters learned how to make a survival hut from watching squirrels make their drey.

Nature and children are natural playmates; they're both wild and messy, unpredictable and beautiful.
—Mark Hoelterhoff (psychologist)

● **Explore Your Special Spot:** Find places to climb and to jump, taking gentle risks. Can you balance on that log? Can you create a small obstacle course and challenge your friends to complete this?

Credit: Anne-Marie Jackson

● **Your Favorite Tree:** Select a favorite tree. How does it change from week to week? If it could speak, what would it say to you? Look for squirrel or bird nests; do any insects live here? Can you climb onto its lower branches?

● **A Quiet Sit:** Stay still in your favorite place; slip your hands behind your ears and really listen. How many sounds can you hear? What different colors do you see? What are the smells of this place? Feel the textures around you, of rocks, of moss, of bark. Can you see clouds from your spot—what shapes do they remind you of?

● **Keeping Track:** Watch and keep a record of the animals that visit your special place (mammals, birds, insects, reptiles, and amphibians); can you find evidence that they were here (tracks, scat, broken branches)? Can you follow an animal trail (look for bent grasses, disturbed leaves, scuffs in the ground, tracks in the mud)?

> There are two gifts we should give children; one is roots and the other is wings.
> —HENRY WARD BEECHER
> (civil rights activist)

● **The Very Small:** Observe the veins of a leaf, the colors of a rock, the particles of soil, the petals of a flower. Using popsicle sticks and yarn, can you create a micro-trail? Poke your popsicle stick into the ground near points of interest. It might be a spider's web and hole in the ground and chewed leaf. Join your popsicle sticks with the yarn, creating a small and colorful trail. Bring your family or friends to your trail and give them a guided tour of your discoveries.

● **Hole Patrol:** Many birds and mammals use cavities or holes in trees for their shelter. Often these are left over from wood-

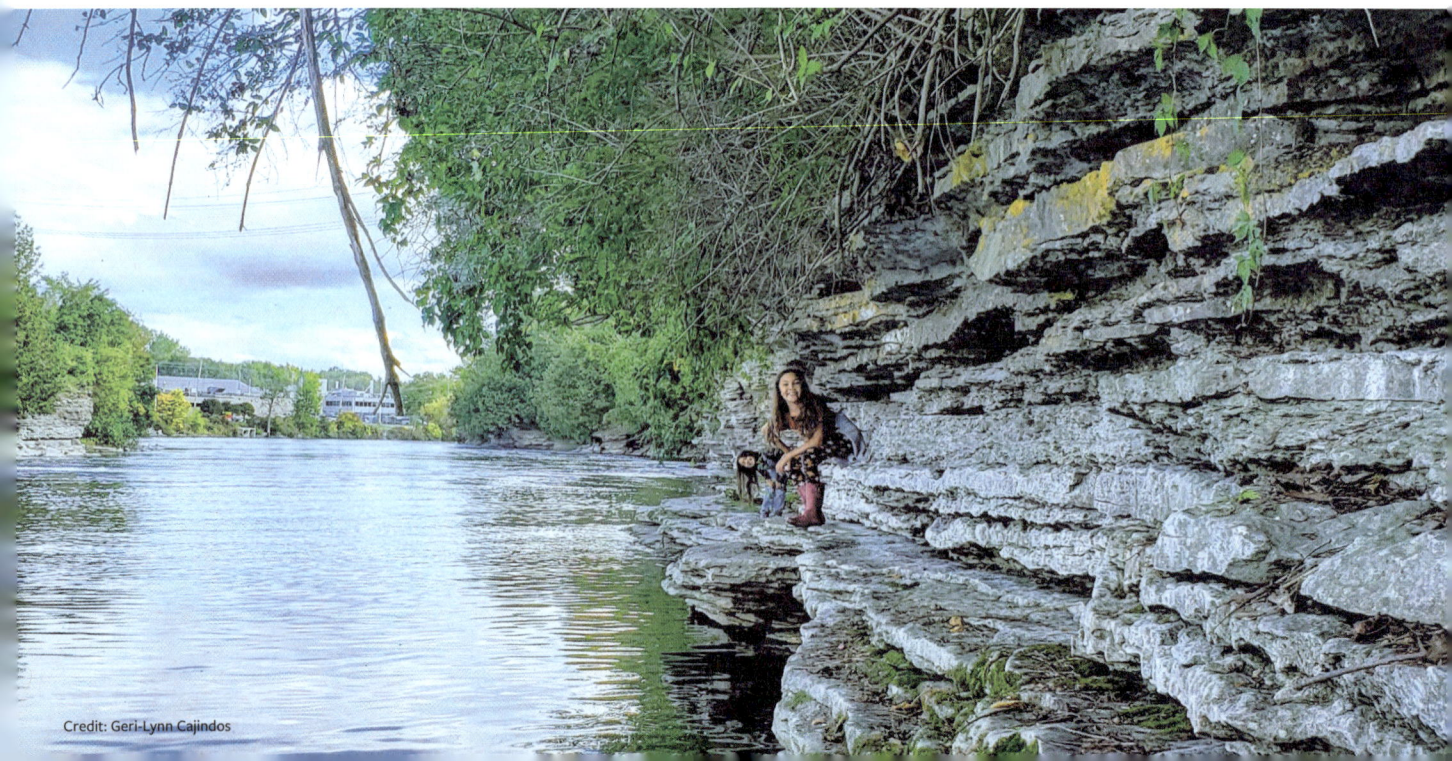

Credit: Geri-Lynn Cajindos

peckers who were looking for insects. They are great hole makers! Old trees often have many holes as do dead trees. Other animals live in holes that they make in the ground. Explore your special spot. How many holes can you find? Who do you think lives there? Can you find any signs that they have been visiting lately?

● **Bird Whispering:** While sitting in your special spot, you may be able to call birds to you! How? By practicing something birders call "pishing." Just repeat the word "pish, pish, pish, pish." Look this up on the internet to hear how to say the word. By repeating this sound over and over again, you'll attract small songbirds like chickadees and nuthatches or other birds that live near you. They'll come in close to where you are sitting because this sound makes them very curious.

● **Create a Secret Trail:** Use arrows and secret signs such as a faint line of pebbles, a crossed stick, a mark in the soil so that you can lead a friend or your family along a mystery trail with a prize or a special treat at the end.

● **Map Making:** Make a colorful map or mural of everything you've seen in your special place, and keep adding to this, every time you visit.

● **Giving Back:** Is there something you could bring or something you could do to help keep your special place healthy? You might consider bringing a little compost from home to sprinkle around the plants; you might bring a few sunflower seeds for some chipmunks, or a small plate of water for birds and toads. Do you have other ideas? Talk about your ideas with an adult.

Credit: Emily Warren

We visit our fort each day! It's a very special place to us, as it provides such wonderful opportunities to problem-solve as we build and engage in dramatic and imaginative play.

—EMILY WARREN
(early childhood educator)

70

AGES 6–7 YEARS
Landmark 10

Plant, tend, and harvest something you can eat (with help from an adult).

Why?

For a child, it truly is a gift to watch a plant grow. To observe a tiny seed sprout, unfurl, and become a seedling and then a mature plan seems nothing short of magic! And eyes widen in amazement when it is time to harvest! Caring for something alive involves thinking and talking about what it needs to be healthy. Can a plant be strong and healthy without any water? Does it need sunlight to grow? What happens when it freezes in the fall?

As Aldo Leopold once said: "The healthiest food is the shortest distance from the earth to your mouth." Fresh food right from the garden is packed full of vitamins and minerals. And if a child grows something, they are much more likely to eat it. The act of gardening is physically active, your hands are in direct contact with the earth. Soil contains microbes and fungi that are beneficial to a developing immune system. Gardening has been shown to boost serotonin, the feel-good hormone that helps kids feel positive, relaxed, and happy.

At this age, children are developing an expanded sense of time and can begin to connect the various steps of planting, tending, and harvesting plants. Caring for a garden and watching all the life it contains helps children directly experience the complex web of life and strengthens their understanding of how all living things are connected.

Climate Change Connections

All of the local food-related Landmarks help to reduce fossil fuel consumption by reducing the distance that food travels from the ground to our plates. Here's an activity to help children understand the concept of food miles: Examine some of the foods in your kitchen or at the grocery store. Make a list of the countries of origin from the tags on a selection of these foods. If you can access a globe or map of the world, find these countries on the map together. Then find the place where you live. How far did that food need to travel to reach you? How do you think it got here? Talk about various modes of transportation, such as airplane, train, truck, and car. Find a food that was grown locally (perhaps in your own backyard!). How far did it need to travel to reach you? Can you work together to eat more locally grown foods at your home? Climate change education is about learning to make healthy choices—not gloom and doom!

How?

● **Make a Garden Together:** It doesn't have to be big—small is beautiful! Find a sunny place for your vegetables to grow big and healthy. Access fresh soil and compost, mix together and create growing beds. If you don't have much space, consider building simple planter boxes. You can even use old pots, recycling boxes, or whatever might hold soil. You can even grow herbs and lettuce indoors by placing a small planter box in a windowsill where there is good light exposure. Young children can help to plant things with large seeds like beans, peas,

At the end of the day, your feet should be dirty, your hair should be dirty and your eyes sparkling!
—SHANTI
(author and wellness instructor)

or pumpkins and place pieces of potato in planting holes.

● **Grow a Pizza Garden:** Choose vegetables and herbs that can be used to make a pizza! Plant tomatoes, basil, oregano, and peppers. Children can help with planting,

Credit: Heather Snowball

Credit: Heather Snowball

Gardens and children need the same things—patience, love and someone who will never give up on them.

—NICOLETTE SOWDER
(creator of Wilder Child)

watering, and harvesting the ingredients. Once the vegetables are ready, use them to make a homemade pizza with the help of an adult.

● **Seed Saving:** After harvesting your garden, show children how to save seeds from plants like sunflowers, beans, or tomatoes. They can dry the seeds and store them for planting next year. This teaches them about the cycle of life in a garden and sustainability.

● **Plant a Rainbow:** Encourage children to plant vegetables and fruits of different colors, such as red tomatoes, orange carrots, yellow peppers, green beans, blueberries, and purple eggplants. They can enjoy watching their "rainbow" grow and later eat a colorful meal they've helped to create.

● **Fruitful Trees:** If space allows, plant a fruit tree like apple, pear, or cherry. While trees take longer to grow, they provide a valuable lesson in patience and long-term care. Children can help with mulching, watering, and eventually harvesting the fruit.

● **Garden Art:** Let children personalize the garden by painting stones, making plant markers, or decorating pots. They can create a "garden guardian" using sticks, old clothes, and a hat, which adds fun and creativity to the gardening process.

● **Help to Water:** Find a small watering can that's not too heavy when filled with water so children can help water the garden. Fill a large bucket with water from the hose and let them dip their watering cans to fill them.

● **Mulching:** Children can help spread mulch of straw or leaves around plants to keep their roots cool and moist; they can gently tuck the mulch around the plants.

● **Hideaways:** Grow vining plants like pole beans or morning glory over a frame or teepee of poles to make a special hiding place for children in the garden.

● **Compost:** Start a compost pile to recycle discarded plants and kitchen scraps when food is harvested. This reminds us to give back to the earth that feeds us!

● **Garden Visitors:** What other living things visit your garden? Do you see holes in some leaves? Who do you think ate them? Can you find any insects, birds, or small animals in your garden? Make a booklet with pictures of "Visitors to my garden."

● **Herb Planter Box:** Your garden can also be a small planter box on your balcony or in your windowsill. Grow herbs such as basil, mint, chives, and rosemary. Give chil-

> **Last weekend we dug up the potatoes that have been growing in our garden all summer. We've grown potatoes for years but this year they were HUGE. The kids were screaming with excitement every time they found a potato. It was the best kind of treasure hunt! Next, we will be pulling out our carrots and beets for a big Thanksgiving feast on the weekend!**
>
> —HEATHER SNOWBALL (teacher)

dren the responsibility of caring for their planter box.

● **Share a Meal:** Prepare some of your vegetables in a meal for your family or friends (ask an adult to help you).

Credit: Sheila Potter

AGES 6–7 YEARS

Landmark 11

Find 3 ways to recognize and enjoy the change of each season.

Why?

The natural world operates in an exquisitely timed rhythm of seasonal events. From the unfurling of leaves in spring to the dried flower heads bursting with seeds in autumn, each season offers its own unique gifts of discovery and wonder. As children grow and their understanding of time deepens, observing and celebrating these seasonal changes strengthens their connection to the world around them. This awareness fosters an appreciation for the distinct activities of each season, creating anticipation and excitement for what comes next. Finding ways to recognize and enjoy the changes of each season enriches children's understanding of the natural world, instills a sense of awe and curiosity, fosters a deep emotional connection to nature, and teaches valuable life lessons about adaptability and resilience.

Climate Change Connections

One of the unfortunate impacts of climate change is more unpredictable weather and a disruption of the normal patterns of seasonal flow. While early elementary aged children are too young to understand the complexities of climate change, they can notice unusual weather patterns, such as extended melts in winter and more frequent storms in summer. One of the best ways we can combat climate change is to explore many ways of having fun that don't require electricity or fossil fuels. Here are some seasonal ideas to get you started.

How?

In Fall...

● **Make a Mushroom Print:** Mushrooms are the fruiting bodies of fungi. Gills from under the head of the mushroom expel thousands of tiny spores. You can make a beautiful print from the spores. Make sure you only do this with a knowledgeable adult. Have the adult who knows their mushrooms well, collect several different nontoxic species. When you are at home or in the classroom, gently remove the stalk so just the mushroom head remains. Place the mushroom heads on a piece of paper, gill side down. Place a few drops of water on each mushroom head. Cover with a glass and leave for a few days. Remove the cover and carefully lift the mushroom straight up; avoid smudging by sliding the head from side to side. On your paper will be a beautiful and delicate spore print. Each species of mushroom leaves a print with a distinctive pattern and color, which can be used to identify the species. Hang up your mushroom art and enjoy!

● **Adopt a Tree:** Have students select a tree. Ask them to name their friend. Try to visit this tree at least once per week throughout the seasons. Find out what kind of tree it is. Use a journal to make a drawing or take a photograph showing how the tree changes throughout the seasons. Use a clothespin to clip onto a leaf just as a it is about to change color. Have a child write their name on the clothespin and watch how quickly the leaf changes color as the days become cooler.

Autumn leaves don't fall, they fly. They take their time and wander on this their only chance to soar.

—Delia Owens, from *Where the Crawdads Sing*

● **Mouse House:** Fill some small containers with hot water (old pill bottles, film cannisters, small yogurt containers). This will represent your "mouse" that is trying to stay warm in the winter. Measure the temperature. Have children make a nest using only natural materials found outside. Think about the kinds of material that might keep your mouse warm. Place your mouse inside

Credit: Emily Warren

Where flowers bloom, so does hope.
—Lady Bird Johnson (former U.S. First Lady, early conservationist)

the nest. Leave for at least an hour. After an hour, rescue your mouse and measure its temperature. How warm did your mouse stay? What materials were used to help keep the mice the warmest?

● **Make Leaf Art:** Collect various shapes, sizes, and colors of leaves. Arrange these and glue them onto paper to create leaf animals or insects. Using a hot iron (parents and teachers only!), seal different colored leaves between wax paper and hang these in your window. These become leaf-stained glass windows!

● **Leaf Patterns Shapes and Sizes:** Collect a variety of shapes, colors, sizes of leaves (from as many different trees as you can). Arrange them from smallest to biggest in a line. Create leaf rainbows by going from the lightest to darkest colors. Use 4 sticks as a quadrant and organize leaves into various categories by shape (long and slender, short and squat, pointy, rounded).

● **Harvest Fun:** Visit a local fall fair, farmer's market or harvest celebration. What kinds of foods are harvested in the fall where you live? Can you make a craft using nearby materials, such as a corn husk doll or a dried flower bouquet?

Winter

● **Tracking:** If you have snow, find some animal tracks and follow them. Ask: What direction was my animal going? What was it doing? What kind of animal made those tracks? Back at the classroom or at home, use a piece of paper to create your own tracking story and ask someone to see if they can figure out what happened.

● **Catch a Snowflake:** Using a laminated piece of black bristol board, make a snow-flake catcher. On a calm day when it is snowing, go outside and catch some snow-flakes as they fall. Many of the most interesting shapes form when the temperature

Credit: Matthew Walmsley

is close to freezing. Make sure your bristol board has time to cool before starting. Can you find stellar crystals, hexagonal plates, needles, or columns?

● **Measure Your Shadow:** Because during the winter, the sun is much lower in the sky, shadows are longer. Go out into the schoolyard or backyard. Stand still and have someone measure your shadow. Record the result. Now do this again in early summer at the same time of day and compare. Depending on where you live, your shadow can be up to two times longer!

● **Indigenous Stories:** Learn about seasonal activities of local Indigenous communities. Winter is storytelling season in many Indigenous cultures.

● **Snow Sculptures:** Build a snow fort, snow person, or snow animal.

● **Winter Sports:** Go skating, skiing, or tobogganing. Can you think of other winter sports to try where you live?

Spring

● **Make a Bird Nest:** Can you make your own bird nest using mud and grass? How well does your creation compare to a real bird nest?

● **Listen to Sap Rise:** If you live in a northern area with maple trees, on the first warm days of spring use a stethoscope to listen for sap rising. Select small maple trees about 10 to 12 inches in diameter (the thinner bark will make it easier to hear). Rising sap will make burbling, crackling, and popping sounds.

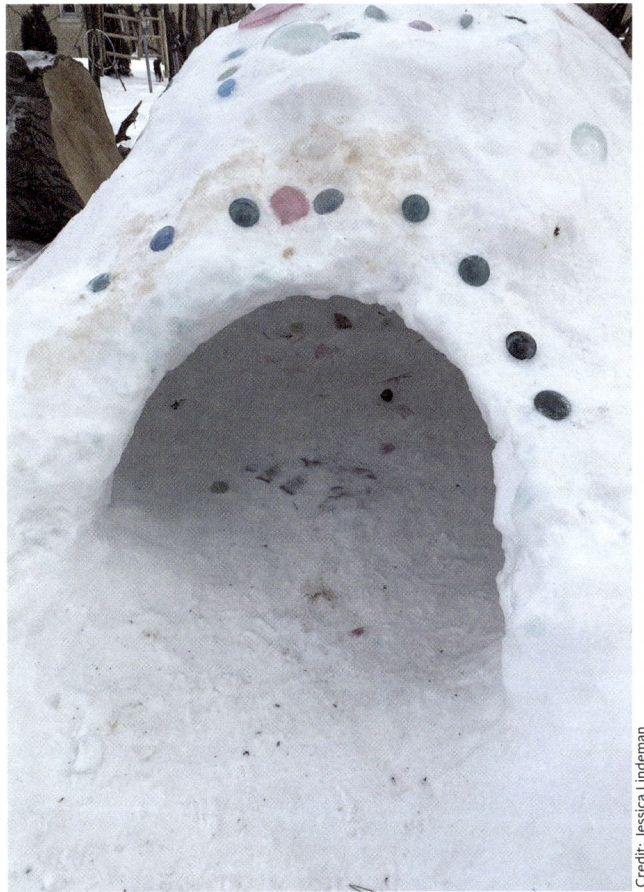

Credit: Jessica Lindeman

● **Clothespin View:** Clip a clothespin onto a bud just before it opens and write your name on this. Come back every day and see how quickly the bud unfurls.

● **Listen for Birdsong:** Using Merlin (a free bird app), identify birds that are singing in your "neighborwood."

● **Green Sheen Challenge:** Using the green color wheel provided (page 195 in Resources), clip small bits of green plants with a clothespin to the best match on the wheel. How many different shades of green can you find in the natural world?

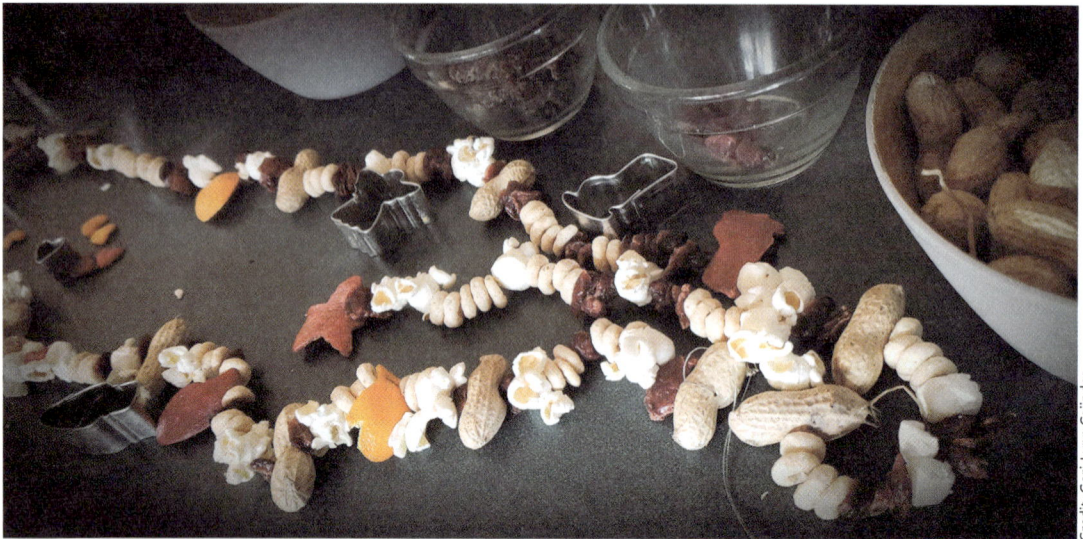

Credit: Geri-Lynn Calindos

Today we welcomed August by enjoying local corn that we husked and cooked ourselves. We used the husks to create lovely little corn husk dolls. What a wonderful seasonal activity that reminds us to waste not, want not! Our creations will provide us with endless hours of imaginary fun and allowed us to practice our tying and weaving skills.

—EMILY WARREN
(early childhood educator)

● **Flower Power:** What are the first flowers to bloom in the spring where you live? Take pictures of different flowers throughout the season, and notice the date when they started to bloom. Continue through the summer to create a seasonal parade of flowers!

Summer

● **Smell Cocktail:** With a wet sponge, dab your upper lip. The moistness of the air helps you to smell better. Fill a small cup full of different scents from a meadow or forest. A few evergreen needles, some soil, a few fragrant flowers, a bit of mint. Use a small stick and crush your concoction to release the odor. Share with others. Give your creation a name—forestopia, flowerlicious, meadowmagic.

● **Catch an Insect:** Lay a white sheet in front of a bush and give the bush a shake. All kinds of invertebrates will fall out. You can also use a net to sweep through the grass. Using tweezers, carefully place what you've found in a container. With the help of Seek from iNaturalist, can you identify your invertebrates? Here are some questions you might ask. Are these adults or larvae/nymphs? Can you observe them breathing? What do they eat? Are they camouflaged? How do they move? Please release all your bugs with thanks.

● **Wildflower Weaving:** Find a forked branch (around 2 to 3 feet in length). Wind raffia or colorful yarn back and forth between the fork. Now weave in different flowers, stalks of grass, and leaves through the strands of yarn. Stick your creation into the ground so it stands upright. Have a weaving exhibition to show off your creations.

● **Berry Tasty:** What kinds of edible berries grow where you live? Can you find strawberries, raspberries, blueberries, or others nearby? Be sure to check with an adult before tasting wild fruits, or visit a nearby pick-your-own farm.

© Roland / Adobe Stock

AGES 6–7 YEARS
Landmark 12

Meet the friends in your "neighborwood." Get to know at least five different kinds of plants, insects, birds, and mammals that live nearby.

Why?

A community is made up of many kinds of living things. Learning to recognize and develop positive relationships with other-than-human beings makes the world a friendlier place and opens the door to learning the stories of these new friends. The Anishinaabe say that when you know the name of a thing, it calls out to you as you pass by. Instead of seeing a "green smear" of plants, different species begin to resolve themselves into unique beings. When you get to know a new friend, the first step is to know your friend's name and the story and the character behind the name. Haudenosaunee scholar David Newhouse says that we live in a storied landscape, and to learn the stories, we must first learn the names of the things that live here. For example, that tree over there is a white pine. You can tell because it has needles in clumps of five, just like the five letters in "white." It is an analgesic, meaning you can make a tea out of the needles that can help reduce pain. The tree is often sculpted into interesting shapes by the prevailing winds. The tree sings its own song, as the breeze strums the branches into a musical whooshing sound. Other trees have their own unique song as well. What other stories can you learn about the trees that live close to you?

Climate Change Connections

Learning the names of the things that live near us means getting up close and personal to have a good look. When we're always whizzing past in a car or on a bus, we don't even notice many of the things we pass by. The world looks very different when you travel on foot, and you can stop to look, listen, smell, and touch the things you're passing. Using the energy in our own bodies to travel whenever we can is not only climate friendly, it keeps us healthy too!

How?

● **Nature Apps:** Use the wonderful free app called Seek. By simply holding your phone over any species—insect, flower, shrub, tree, or bird—the app will tell you what you are looking at. The data gets reported to citizen scientists. You can also use Merlin which will help you identify any birdsong and calls that you hear. Remember that a name is only a beginning and that there is

a fascinating story behind every name. See if you can find out at least three interesting facts for every species you find. Focus on native species that evolved in your area if you can.

● **Play Tree Tag.** Go to an area with at least four different tree species. Show the children the distinguishing characteristics (e.g., leaves, bark, shape) of the various trees. Designate a person as It. It shouts out the name of a tree, e.g., "White Oak!" The children need to run and touch the branch or trunk of the correct species to be "safe." Anyone tagged or who runs to the wrong tree becomes It. You can also shout out "coniferous/deciduous" or "opposite branching/alternate branching," or "wavy edged leaves/smooth edged leaves."

● **Basement Windows:** Much of what goes on in the world is hidden from view. You can open a window onto these special places by simply lifting up a rock or a log

> **Whoever you are, no matter how lonely, the world offers itself to your imagination, calls to you like the wild geese, harsh and exciting— over and over announcing your place in the family of things.**
>
> —MARY OLIVER (author)

and peeking underneath. You might be lucky enough to spot a salamander, or a plump earthworm, or a lumbering millipede. What magical creatures can you find? What are they and how do they help to maintain a healthy and diverse forest ecosystem? Don't forget to carefully close your window when you are finished.

● **Set Up a Window Birdfeeder:** A stick-on birdfeeder that attaches to your classroom or home window is a simple way to attract birds and gives you the opportunity to study them up close. Select a window that is near some bushes. Birds always feel more comfortable if there is nearby cover. Fill your feeder with black oil sunflower seeds. Be patient and birds will come. Observe how they fly, how they feed, and how they interact with one another. You can also periodically remove the feeder and, standing next to the feeder, fill your own hands with seeds. Soon small birds such

> **To be without trees would, in the most literal way, mean to be without roots.**
> —RICHARD MABEY (author)

as chickadees and nuthatches will learn to feed right from your hand! Frank Glew calls the gentle touch of a bird's claws on your palm "that chickadee feeling," in his lovely children's book of the same name.

● **Go on a Sock Walk:** It is an interesting fact that while plants can't move, they still are able to spread far and wide. Of course they do this by dispersing their seeds. How seeds travel is fun to explore. Hand out a large sock to each person and have them cover one shoe. Walk through the long grass of a field. Examine your socks. How many hitchhikers stuck on for a ride? Use a hand lens to study how the seed was able to stick to your sock. What kind of plants are they from? Use an egg carton to collect different seeds and sort them by how they travel. Some seeds sail on the wind (dande-

lion, milkweed, goat's beard), some twirl (maple, ash, even the seeds inside cones!), some pass through the digestive tracks of animals (berries, nuts), and some even explode (jewelweed, violets).

● **Visit a Local Pond, Stream, or Wetland to Find Out Who Lives There:** All you need are a few nets, a white bucket, tweezers, some rubber boots, and if you have one, a magnifying glass. Make sure the kids are closely supervised. Sweep your net in the pond or stream, scooping up gently among plant roots, near rocks, or in the mud. Carefully sift through your findings. Soon enough, you'll spot something wriggling. Using the tweezers, gently remove the wriggler and rinse it in clean water and place this in the white bucket. Continue and repeat. What did you find? Often these are the larvae or nymph stage of an adult such as a mayfly, dragonfly, or stonefly. In the evening, cup your hands behind your ears and listen to the sounds of a wetland symphony. In the spring and through the early part of the summer, frogs sing to call for a mate and to establish territory. Each species has a distinctive call that you can identify with the help of apps such as Frog-Watch. Frogs are bioindicators meaning that a healthy frog population signifies a healthy ecosystem.

● **Nature Journal:** Keep a nature journal of your findings. The more you recognize the more you'll feel part of the bigger life systems that nourish us all!

● **Measure the Height of a Tree:** Find a tall tree. Now find a stick that is as long as your arm. Hold the stick straight up in your hand and make sure your arm is horizontal to the ground. You've made a perfect right angle with the stick and your arm. Close one eye and carefully walk backward (make sure to look behind you as you do so; please don't trip or walk onto the road!). Walk until, through your one open eye, the base of the stick is even with the base of tree and the top of the stick is even with the top of the tree. Now from where you are standing, count how many steps it takes you to get to the tree. That is how many steps your tree is tall! If you want to convert that number to yards, measure how long is each step you take while walking at a natural pace (roughly ½ or ¾ of a yard). Multiply that fraction by the number of steps it took to walk to the base of the tree. That's the tree's height in yards! How tall is your tree?

● **Age a Pine Tree:** If you come across a pine or spruce tree, you can roughly calculate its age by counting the whorls (each year these trees grow, they send out a whorl of branches which are like the spokes of a wheel). Each whorl represents a year. The spaces between the whorls indicate what kind of growing season there was in that year. A big distance between whorls means there was abundant rain and sunshine; a short distance means a challenging growing season. To age your tree, start at 3 (since it took that long for a tree to begin growing whorls), and count how many whorls you can spot. That gives you a rough idea of the age of the tree. Don't forget to look for stubs on the lower part of the tree because evergreen trees slough off their lower branches when they can't access sunlight. How old is your tree?

We visit our school nature area at least twice each week, and take time to play in the nature playground and discover at least one new plant species each visit using the app Seek. We have so much fun trying to find the plants we have previously identified and make it a game to try and remember all the species names. Some of our students started the year feeling very uneasy going into the nature area as they had not had many outdoor experiences in a forest setting. There has been a complete shift and after learning about the plants in our forest, and feeling a sense of connection to the area, that is no longer a concern for us!

—HOLLY PODRES-BEATON (teacher)

Credit: Heather Snowball

Landmarks for Age 8 to 9 Years

Age 8 to 9 Years: Characteristics of This Age Group

At ages 8 to 9, children are now capable of grasping more complex concepts and are eager to understand the "why" of things. They might be interested in why birds fly south for the winter or why a woodpecker doesn't damage its head by slamming its beak into solid wood.

Socially, 8- and 9-year-olds are becoming more collaborative and enjoy working in groups. They are just beginning to understand the power of collective action, making them receptive to team-based environmental projects. Whether it's starting a pollinator garden, or helping with a school-wide recycling program, they are motivated by the idea of contributing to a common goal with their peers.

This is an age where play and imagination are one of the primary ways that children engage with their immediate environment, especially with the natural world. For young children, their imagination can transform natural settings into magical worlds, where a cluster of trees becomes a castle, or a hollow in the ground turns into a dragon's lair. Imaginative play allows children to project their emotions and ideas onto the natural world, creating personal stories and connections. At the same time, children of this age love to engage in physical play, and by doing so, children develop gross motor skills and become more confident and comfortable in outdoor settings. Activities like running, jumping, or navigating uneven terrain strengthen their bodies and increase their resilience.

At this age, children demonstrate a heightened sense of justice and fairness. They are becoming more attuned to issues of right and wrong, particularly when it comes to how they, or their friends, are treated and how we are treating the environment. They may feel strongly about protecting animals, enhancing habitats, or preventing pollution and are eager to voice their opinions and advocate for change.

A school principal told a wonderful story about this age group. Some of the grade 3 and 4 students noticed and were worried about a bird with a broken wing, scurrying about their school's gravel parking lot. Even though he recognized the bird, he fetched a field guide and encouraged the students to identify the bird. It was a killdeer, a plover with black bands about its neck. Upon further research, they discovered that sometimes a killdeer will pretend to have a broken wing, to lead predators away from its nest. That prompted them to look around, and sure enough, they discovered a hollow dip in the gravel: a killdeer nest containing four beautifully camouflaged speckled eggs. The students suddenly became worried that the eggs would be crushed by cars; so with the principal's help, they erected a barrier using flagging tape. Over the following weeks, they protected the nest, educated other students about killdeers, and using binoculars kindly provided by the principal, they watched Mama Killdeer brood her eggs. They watched the young hatch and quickly run about. By this time, these students were bound and determined that nothing would happen to "their killdeer" and its young. That killdeer revealed herself to the students, not just as any bird but as a singular living being with a character and a story that called out to each one of them.

This story illustrates how children at this age can begin to understand the importance of taking on responsibility and the power of making personal connections with the natural world.

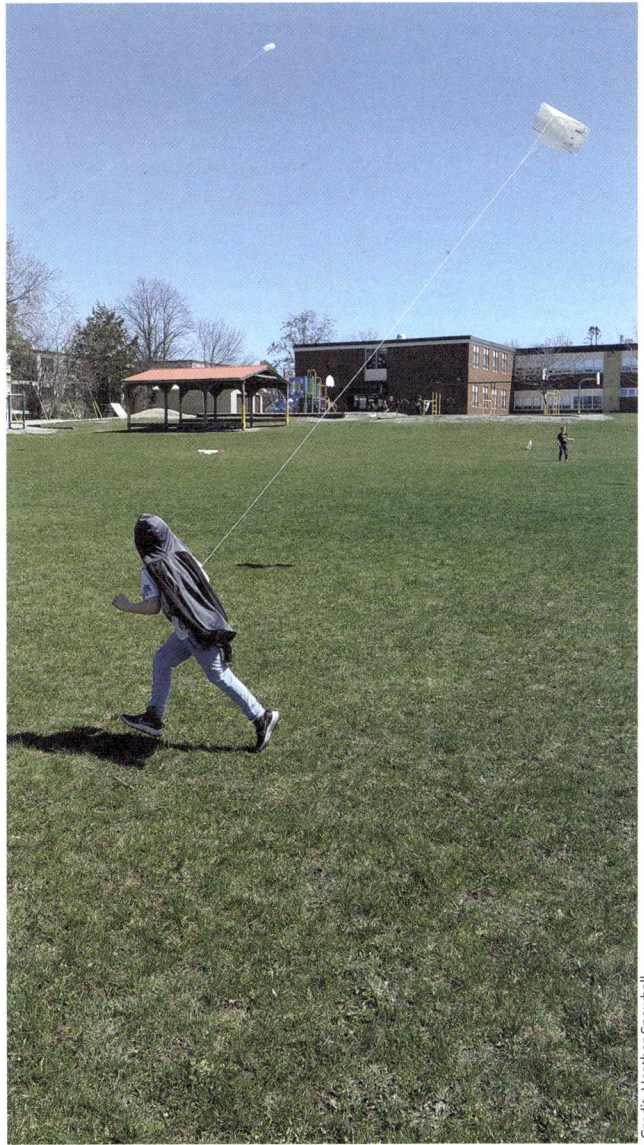

Credit: Heather Snowball

AGES 8–9 YEARS

Landmark 13

Travel a familiar route regularly by yourself or with a friend. This can include walking, riding your bike, or travelling on public transit.

Why?

Children notice so much more about the world around them when they make the decisions on how to navigate a route themselves. And, by noticing details in familiar places, children deepen their sense of connection to places close to home. Having opportunities to travel independently contributes to self-esteem and strengthens leadership and problem-solving skills. While some parents may be hesitant to let a child travel on their own, going with a friend can be helpful, or walking the family dog with a cell phone can be good safety options. Educators can help school classes develop their navigation skills by working with the class to plan a route together and have students take turns leading the class to travel in familiar places. Learning to travel independently literally opens a door to exploring the world. And, that is a gift that will last a lifetime!

Climate Change Connections

For many of us, automobiles have become our primary mode of transportation. Our collective addiction to gasoline-guzzling cars is a major contributor to the global climate emergency. Encouraging children to be comfortable walking, cycling, or travelling by transit has tangible environmental benefits, as well as benefitting their own physical and mental health.

How?

● **A Word About Safety:** If you are worried about safety, try reconsidering actual risk. Studies have shown that children are much more likely to be injured as a passenger in a car than by walking or biking.

● **Independent Mobility:** Try walking all or part of the way to school alone or with friends. If adults go along too, let the children decide the route and navigate themselves.

● **Keep a Nature Journal:** Record the types of living things you see on your route in your journal. What birds do you see or hear? Are there big, mature trees? Are there young tree saplings? Do you notice any mammals along the way? What changes do you notice as the seasons change? Use Seek from iNaturalist to help you identify nature along the way.

● **Routes:** As a class or a family, find several different routes to the same place. Let the children select the route for today. Which was the prettiest, which was the fastest, the one with the least traffic?

● **Maps:** Travel a route silently with a friend or small group. When you return to home or school, make a map of everything you remember. Later, travel the same route again and see if you can add more details to your map.

● **Tree Faces:** The human brain is very good at seeing "faces," even when they are not really there. Look for "faces" in old trees you walk past. Look for eyes, a nose, and a mouth. The face you see may only have one or two of these features but will still look strangely human. Old willow trees are a good choice for this activity. What might the personality of your tree be? How many faces can you discover?

● **Shapes and Signs:** Urban landscapes are filled with shapes. Can you find circles, squares, rectangles, or 3D shapes (cylinder, cube, cone)?

● **Walk:** Can you find all the letters in the alphabet in the shapes you discover on your walk? A branch might have a V, a U, or an X. A mark in a stone might be an O or an I. Take a picture of each and create an alphabet that you can display.

● **Ant Challenge:** If you spot some ants along the way, crouch down and observe them. Where are they going? Take a picture using the app Seek and try to identify the species. If there are a line of ants, take your finger and rub in a straight line across the trail that the ants are walking. What happens when an ant reaches your rubbed line? Ants use pheromones or smells to follow one another. Your finger mark interrupts

> **If we were meant to stay in one place, we'd have roots instead of feet.**
> —Rachel Wolchin (author)

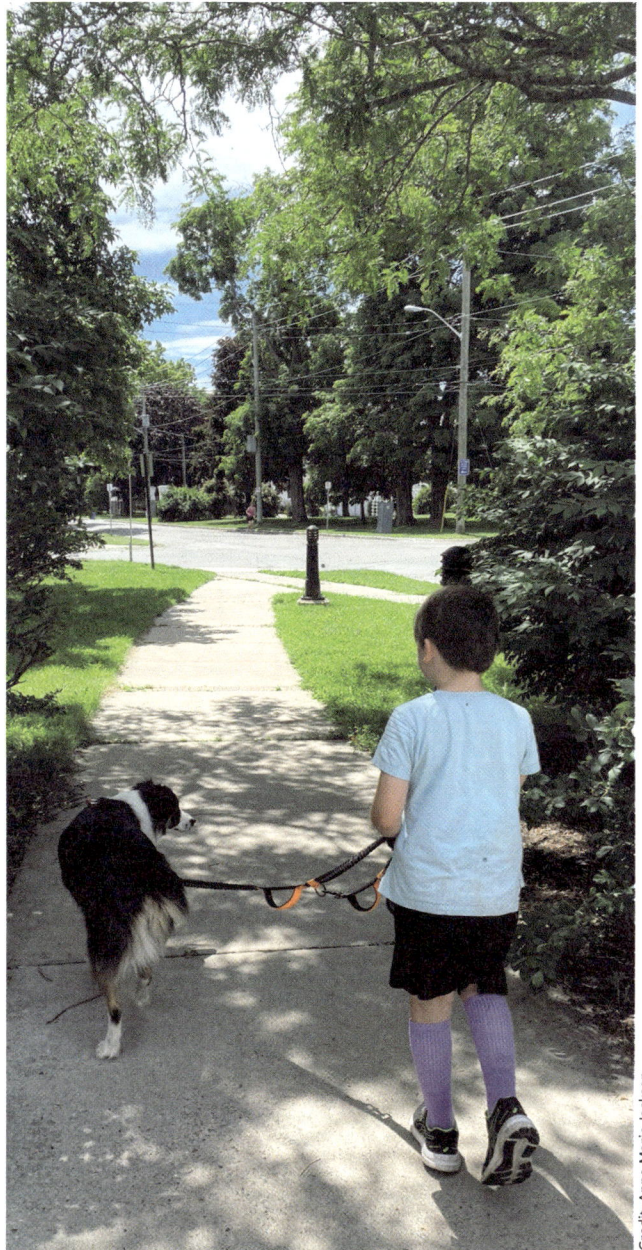

Credit: Anne-Marie Jackson

> **Today we went for a neighborhood walk. After, we looked at a map surrounding our school, traced our path and tried to remember the street names that we observed on our walks and the way the streets curved.**
>
> —MANDY DUFRESNE (teacher)

the scent trail. See how long it takes for the ant to rediscover the trail.

● **Mini Challenges:** Give yourself or a friend a challenge along the way, like counting how many EV cars you see, spotting all the red flowers, or noticing how many people are walking dogs. These small challenges make the trip more engaging and help you pay closer attention to your surroundings.

● **Seasonal Photography:** Take photos along your route during each season—spring, summer, fall, and winter. Compare how the landscape changes over time. Create a collage or scrapbook of your seasonal journey.

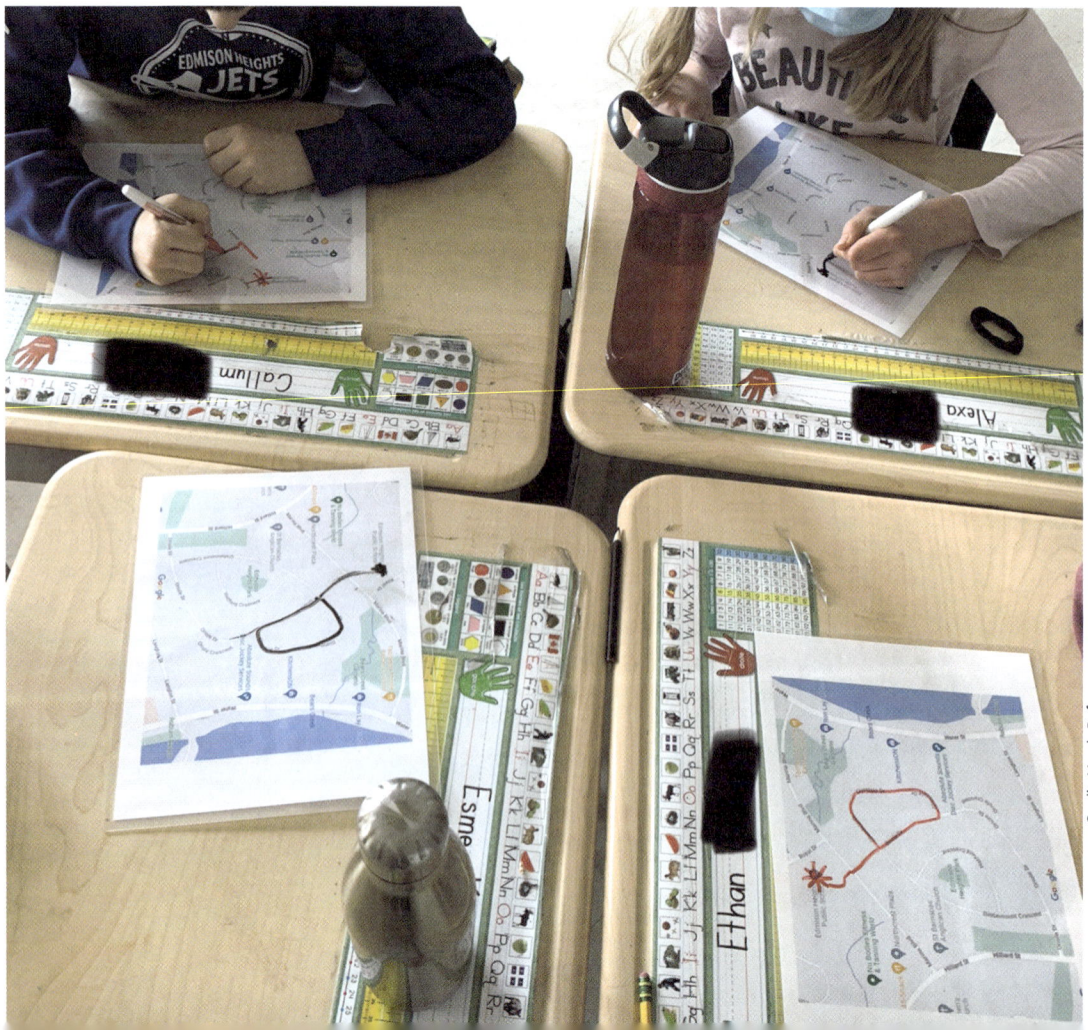

Credit: Mandy Dufresne

Credit: Brooke Nancekivell

● **Friendship Tree:** Choose a tree along your route to visit regularly. Track its growth and changes throughout the year. Give it a name and note the differences you see each time you pass by, such as new leaves, flowers, or fallen branches.

● **Story Route:** Create a story inspired by things you see along your route. Imagine the trees, buildings, or even animals as characters in your tale. Share your story with a friend or write it down to revisit later.

● **Scavenger Hunts and Bingos:** You'll discover many nature scavenger hunts online. Many are organized as bingos. See if you can find a line of items across, down, diagonally, or all the items on the sheet. We've provided a few samples for you in this book. See pages 196–199 in Resources. Try making your own!

AGES 8–9 YEARS

Landmark 14

Try at least five different kinds of outdoor recreation that don't require gasoline or electricity.

Why?

Children at this age are exercising more independence and are eager to try new things and challenge themselves. This is a perfect time of life to develop habits for physical and mental health and to try new types of outdoor recreation. Children can also begin to understand the factors that contribute to climate change, and to enjoy themselves in ways that don't depend on fossil fuels or electricity. Studies have shown that children who engage in active outdoor play and outdoor recreation not only build their physical fitness but experience less anger, aggression, and have improved impulse control and resilience. And, those activities spent in nature help to reduce stress and depression, increase focus, and reduce ADHD symptoms.

Climate Change Connections

Reshaping our oil-based cultures toward more climate-friendly options continues to be a big challenge around the world. We're all constantly lured by media toward bigger, faster engines and increasing time spent online and with the latest power gadgets. Luckily, adults as well as children can relearn and enjoy a wide range of skills that aren't driven by fossil fuels. This can become a fun family challenge to turn off the screens and engines and have fun outdoors in new and creative ways. Children aged 8 to 9 can benefit from honing their observation and coordination skills with a variety of games such as badminton, skipping, hopscotch, baseball, as well as nature-based exploration—all of which are climate-friendly fun!

How?

● **Biking:** Ride bicycles on nearby trails. Some communities have Learn to Bike programs and will provide you with free lessons and loans of bicycles. Biking is a marvelous way to refine balance, build fitness, and explore the areas where you live. Bike as a family or with a friend. Look for trails that traverse more natural areas.

● **Boating:** If there is water nearby, try boating. A good way to start is with a paddleboat. These are sturdy, easy to maneuver, and are very safe, even if you aren't comfortable around water. There is so much to explore via a canoe or kayak. Take lessons and discover for yourself the beauty of travelling over water without a noisy engine!

● **Snow Fun:** If you are in an area that has snow, try tobogganing, snowshoeing, or skiing. Build a quinzhee—or survival snow shelter. Use a shovel to pile snow in a large mound as tall as you are. Take sticks that are about one foot long and poke these into the mound all the way around. Leave to sinter (or settle) for a day or so. The next day, hollow out your quinzhee. When you come cross the end of a stick, you know you've hollowed out far enough. Make sure to create 3 or 4 breathing holes as large as your fist along the bottom to allow fresh air to come in.

No barefooted, tree climbing, mud pie baking, cloud spotting, puddle stomping, bird calling, wild foraging, star gazing, firefly chasing, den building, stream paddling, rock hunting moment with Mother Nature is ever wasted.

—Nicolette Sowder (creator of Wilder Child)

Credit: Sarah Taylor

This week our class built kites from a Kites in the Classroom kit. Yesterday each student got to decorate their own kite, and we assembled them this morning. The perfect breeze started up by the afternoon, so we quickly headed outside to test them out! The class stayed out for over an hour flying them and would have kept flying for much longer had it not been time to go home!

—Heather Snowball (teacher)

● **Fly a Kite:** Using tissue paper, dowels, and glue, design and build your own kite. Use a tail to stabilize your kite. Try different shapes: diagonal, box, rectangular. Which shape flies the best?

● **Geocaching:** Who doesn't like a treasure hunt? Participants can download an app on their phone to find containers, known as "geocaches" or "caches." Typically, a "cache" is a small container that contains

Credit: Nancy Thomson

a logbook. Geocachers sign the book with their code name, enter the date, and then place the cache back exactly where they found it.

● **Naturalist Clubs:** Join a young naturalist club near you. You'll be mentored by experienced naturalists who can tell you more about the birds, butterflies, plants, and other wildlife that is in your region. There are often birding clubs as well. This is a wonderful opportunity to learn about the songs, habits, and species of birds in your area.

● **Stargazing:** There are many astronomy clubs throughout the United States and Canada. They'll often have telescopes set up and people who know how to use them so that you can get amazing glimpses of the night sky. Download a free app called SkyView. You simply point your phone to a nearby planet or star, and it will tell you what it is! Use this app to find out what planets are visible, what constellations are in the night sky at different times of year, and what interesting astronomical events are happening.

● **Hiking:** Together with your family or class, organize a nature hike. Almost every community has a conservation area, a park, or a land trust somewhere nearby. Pack a picnic and explore the different habitats found there. What kinds of wildlife did you see? Use the app Seek to help you identify some of the different species you encounter.

Credit: Shannon Cannon

AGES 8–9 YEARS

Landmark 15

Habitat Explorer: Try these activities—grow a wildlife garden, set up a birdfeeder, catch insects, go fishing, get to know a habitat.

Why?

This is an ideal age for deeply exploring the world and discovering your identity in the miraculous web of life. As Chief Seattle is credited with saying (paraphrased), "Humankind did not weave the web of life, they are just a strand in it." Children begin to recognize their power to influence the world around them and the importance of not just taking from nature but also giving back. Beyond simply reducing their harm, they can actively do something good for the environment. Bringing nature back to built spaces is one way they can accomplish this. Each positive act such as rewilding a backyard or schoolyard, is a seed of "agency"—knowing that you have the ability to act, to make a difference in ways that benefit the well-being of yourself and others. And in the end, that is at the very heart of stewardship.

Climate Change Connections

Often, media presents environmental challenges as overwhelming, scary, and hopeless. This only tends to repel or paralyze people, which does nothing to address or solve the problems. Every child deserves the opportunity to experience the wonders of the natural world, and to understand that each of us can have a positive impact on the world. When we work together toward a common cause, the results can be awesome! Children can be encouraged to become environmental champions by starting with small, manageable projects close to home.

How?

● **Leaf Composter:** In the autumn, collect some fallen tree leaves and pile them in a small enclosure (a loop of chicken wire works well). Watch how the pile gradually gets smaller and starts to look like soil. In less than a year, you should have some lovely leaf compost to sprinkle in your gardens or under your trees to help them thrive. We need to give back to the earth that feeds us!

● **Wildlife Garden:** Animals need to eat too. Why not create a small wildlife garden? One great way to start is to create a pollinator garden. North America is home to over 3,600 bee species, of which 90 percent are solitary. Bees and insects of all kinds are wonderful pollinators. For every three bites we eat, one is courtesy of a pollinating insect. Here are some plants that will attract pollinators throughout the seasons. Check to see what grows in your region. To attract spring pollinators, plant wild strawberry, wild geranium, wild apple, blueberry, chokecherry, serviceberry, staghorn sumac, dogwood, willow, and lilac. To attract summer pollinators, plant bergamot, buttonbush, purple coneflower, Culver's root, meadowsweet, New Jersey tea, common milkweed, butterfly milkweed, hyssop, Joe-Pye weed, common sunflower, pumpkin/squash, lamb's ear, comfrey, Russian sage, nodding onion, prairie clover, native roses, fireweed. For late summer and fall pollinators, plant aster, goldenrod, blue vervain, cup plant, false sunflower, Mexican sunflower, and butterfly bush (Buddleia).

● **Bird-Friendly Garden:** You can attract birds as well. To attract hummingbirds, plant columbines, iris, salvia, and bergamot. If you'd like a wide variety of birds to

Children grow up hearing how broken the environment is, how broken beyond repair. Plant strawberries together, make wild medicines, paint the sunrise. Show them proof that for every act of destruction, they can sow a seed, however small, of beauty.

—Nicolette Sowder
(creator of Wilder Child)

We are on Earth to take care of life. We are on Earth to take care of each other.

—Xiye Bastida (climate activist)

come to your garden, think about planting shrubs such as serviceberry or elderberry. Also plant sunflowers, coneflowers, mulberries, crab apples, wild grape vine. If you can, put up bird feeders, bird houses, and

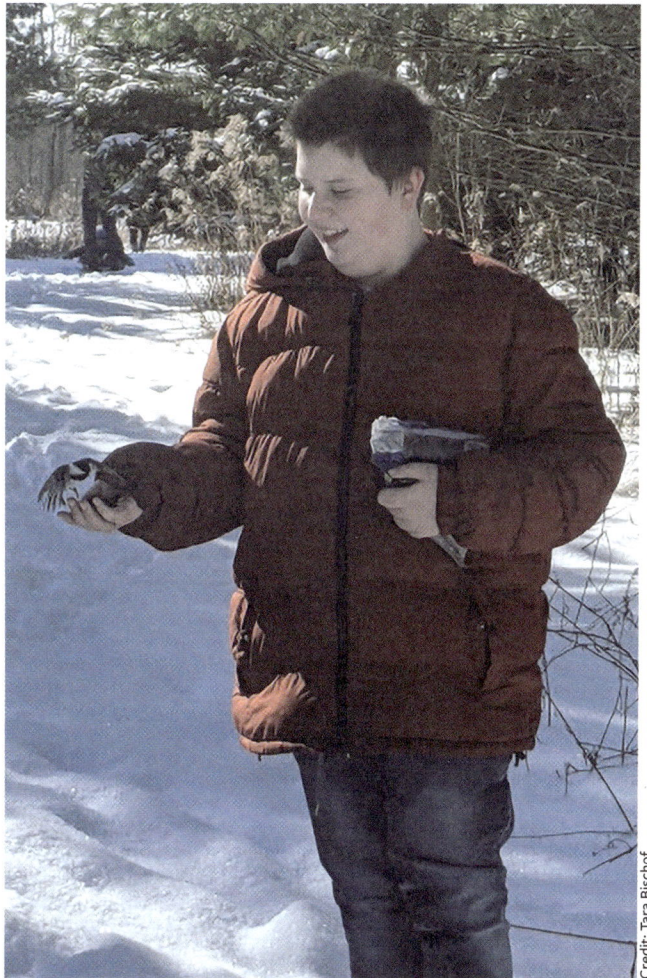

Credit: Tara Bischof

bird baths. Don't forget to use apps such as Seek to find out who is visiting your garden. Keep a small journal of who visits during each season of the year.

- **Birdfeeders:** Set up a birdfeeder at home or at school; watch it every day. Keep a record of who visits, what you see when they visit, how often they come, what time of day, weather, etc. How many different kinds of birds visit your feeder?

- **Sensory Garden:** Why not create a garden with wonderful smells, textures, and taste? Plants such as lemon balm, anise, lovage, basil, and mint all have pleasant smells. Fuzzy plants such as celosia, lamb's ear, hens and chicks, and strawflower have amazing textures that are safe to touch. And of course, there are lots of edibles to plant too—berries such as strawberries, blueberries, and gooseberries or savory tastes such as shallots, cilantro, rosemary, and thyme. There may be a gardening club in your community that could offer up suggestions on what kinds of plants would work best for a sensory garden in your region.

- **Fishing:** Go fishing together. What kinds of fish do you catch? Do you find different kinds of fish in different locations (streams, rivers, lakes)? Remember to release the small ones so they can live and mature. What other kinds of life do you notice while you're fishing? All living things need water, and waterways are magnets for wildlife! You'll see more wildlife if you are still, quiet, and patient. Don't forget to

Credit: Dana Geen

watch for turtles, frogs, crayfish, or other aquatic life while you're there.

● **Hula Challenge:** How many different living things can you find in a small area? Toss a hula hoop on a mown lawn and count all the life within the hoop. Now place the hula hoop in a naturalized area. Count all the life you see. Use the iNaturalist or Seek apps to help identify what you find. Notice the difference between the two areas? What are the consequences for creating healthy ecosystems if most of our school grounds and backyards consist of only lawn?

● **Catch and Identify Bugs!** How many different insects can you find in a single day? Take a photo of each before you release them and see if you can identify each one. Try again in different locations and at different times of year.

We have been so interested in the bugs emerging as the weather changes, so we spent the afternoon on a bug hunt! We discovered worms, ants, beetles, and even a firefly larva! We've been thinking about what kind of habitats each different bug might need and are looking into setting up a terrarium where we can watch some plants grow and worms dig!
—**CAT TRITES** (early childhood educator)

Credit: Matthew Walmsley

● **Build a Food Web:** Do some research on what plants, animals, insects (flying, above and below ground), amphibians, birds, and reptile species live in your area. Print out their names in large font. Stick on a picture of the sun in the middle of an empty wall. Then place the names of your local species randomly around the sun. Using yarn, begin to connect all the species that have a relationship. For example, sun to white pine tree to grey squirrel to red-tailed hawk to earthworm (when the hawk dies) back to white pine (grows out of the soil). By the time you are finished, you should have a complex and interconnected web.

Landmarks for Age 10 to 11 Years

Age 10 to 11 Years: Characteristics of This Age

At this age, children become more aware of their own feelings and start shifting from an inward focus to a broader, outward view of the world. They begin to develop a deeper understanding of empathy, communication, and morality, moving beyond simple rules to think critically about what is right or wrong. These social skills are largely shaped by how parents, teachers, and mentors guide and model behavior. The way they learn to perceive and interact with the world now lays the foundation for their belief systems that they may carry with them for the rest of their lives.

Children aged 10 to 11 are natural explorers. They are inquisitive, curious with a fresh perspective that makes the world around them feel new, alive, and full of

Credit: Matthew Walmsley

wonder. Whether it's tracing a faint animal trail through long grass or uncovering life beneath a log, children at this age have an inborn desire to discover. Children also love to collect things such as shells, rocks, or special nature treasures. They continue to thrive when given the freedom to engage in unstructured play, especially in natural settings.

Providing them with hands-on experiences like gardening, exploring diverse habitats, or participating in local conservation projects, such as planting trees or rewilding a fence line, empowers them by making their impact visible and tangible. These activities give children a sense of accomplishment, which strengthens their commitment to caring for the environment. When they see the direct results of their efforts, it deepens their dedication to stewardship, turning their natural curiosity into a force that they realize can create positive change.

© Infinity Lens / Adobe Stock

AGES 10–11 YEARS
Landmark 16

Visit a place that uses renewable energy and investigate how it operates.

Why?

We live in a time where our energy choices are the major cause of climate change. Young people can begin to explore this environmental challenge and learn about energy options that don't rely on fossil fuels. They can directly experience the power of the sun through simple experiments. Students can visit alternative energy sites within their community. Focusing on ways to reduce carbon at home and at school can contribute to their growing sense of agency and understanding the importance of making wise lifestyle choices.

Climate Change Connections

This is an ideal age to learn about the three fossil fuels: coal, crude oil, and natural gas. Brainstorm the many ways we use these fuels. Why are they not renewable? How do they change the Earth's atmosphere when we burn them? This is an exciting time for all the new developments in renewable energy industries, and the potential for developing a climate-friendly economy. It's not easy to change our habits, but exploring the basics of renewable energy can inspire young people to participate in transforming our communities toward cleaner energy sources.

How?

● **Alternative Energy Walk and Tour:** Go for a walk in your community and see if you can find any buildings with solar panels or any evidence of nearby wind turbines. Do you have a hydroelectric station or solar farm in your community? Does anyone in your community have a passive solar house or a house that is off-grid? If you can, see if you can arrange for a tour. Take a poll in your class; do any students live in homes with geothermal systems or have they installed a heat pump? Invite someone to speak to your class about the energy efficient ways of heating.

● **Where Does Your Energy Come From?** Contact your local utilities and find out where the energy you use comes from. How do you travel to school, and what is the source of that energy?

● **Pinwheel:** Make a simple wind pinwheel. Cut a sheet of paper into a square. Fold the square paper diagonally in half both ways to create creases from corner to corner. Cut along the creases, stopping about halfway to the center. Bring the corners to the center, overlapping them slightly. Push a pin through the center where the corners meet. Attach the pin to the eraser end of a pencil. Test your pinwheel in the wind! How fast does the wind spin your pinwheel? From what direction is the wind coming?

● **Solar Oven:** Make a pizza solar oven. Here is how: Clean the pizza box. Cut a flap in the lid with a box knife, leaving a 1-inch border all around the edges so the lid can still close. Fold the flap so it stands

What you do makes a difference, and you have to decide what kind of difference you want to make.
—Dr. Jane Goodall
(zoologist and primatologist)

up when the lid is closed. Cover the inner side of the flap with aluminum foil to reflect sunlight. Create an airtight window over the hole you made in the lid with clear

Credit: Karen Brown

Students explored different forms of renewable energy (solar, wind, geothermal, and hydroelectric) and drew diagrams to show how they work. We planned a virtual visit to Camp Kawartha Environment Center and learned that it uses solar hot water, geothermal heating, conservation, and photovoltaic panels!

—AMANDA HIPGRAVE (teacher)

plastic wrap. Line the bottom of the box with black construction paper to absorb heat. Insulate the oven with rolled-up newspaper along the bottom, ensuring the lid can still close tightly. Set up the solar oven on a sunny day between 11 am and 3 pm when the sun is overhead. Place it in a sunny spot and adjust the flap to reflect maximum sunlight onto the plastic-covered window. Prop the flap with a ruler and angle the box using a rolled-up towel if needed. You can bake nachos, s'mores, cookies, and more!

● **Turbine:** The energy from falling water has been used throughout the world in mills that grind grain or saw wood using waterwheels. Water can also be used to create electricity by turning turbines in hydro-electric generating stations. Build a simple turbine using materials such as paper or aluminum plates, toilet paper rolls, pieces of plastic dairy containers, pencils (as an axle), or other ideas of your own. Can you build something that will turn when placed under a stream of water from a tap or poured from a jug? Find pictures of waterwheels to give you ideas.

● **Solar Water Heater:** To make a solar water heater, use black or dark green garden hose. Arrange the hose in curves to maximize the surface area. Leave the coiled hose in the sun. On a sunny day, run water through the hose—it will come out warm to hot! Fill up a wading pool with the warm water (add cold water if you need to) and enjoy a good soak!

● **Other Options:** Explore other options for reducing our use of fossil fuels, such as renewable fibers for clothing, locally produced food, and active transportation.

● **Explore Battery Storage:** Investigate how renewable energy can be stored using batteries. Students can learn about the role of batteries in storing solar or wind energy for later use. As a project, they could build a simple battery using household materials like lemons or potatoes to light an LED.

● **Investigate Bioenergy:** Explore how organic materials can be used to produce energy. Students can learn about the process of composting and anaerobic digestion. Consider a visit to a local facility that converts organic waste into energy or biofuels, or have students experiment with making their own small biodigester.

● **Community Energy Project:** Organize a project where students help design a community-based renewable energy initiative. This could include proposing a solar panel installation for a community center, starting a composting program, or creating a plan for a neighborhood bike-sharing program.

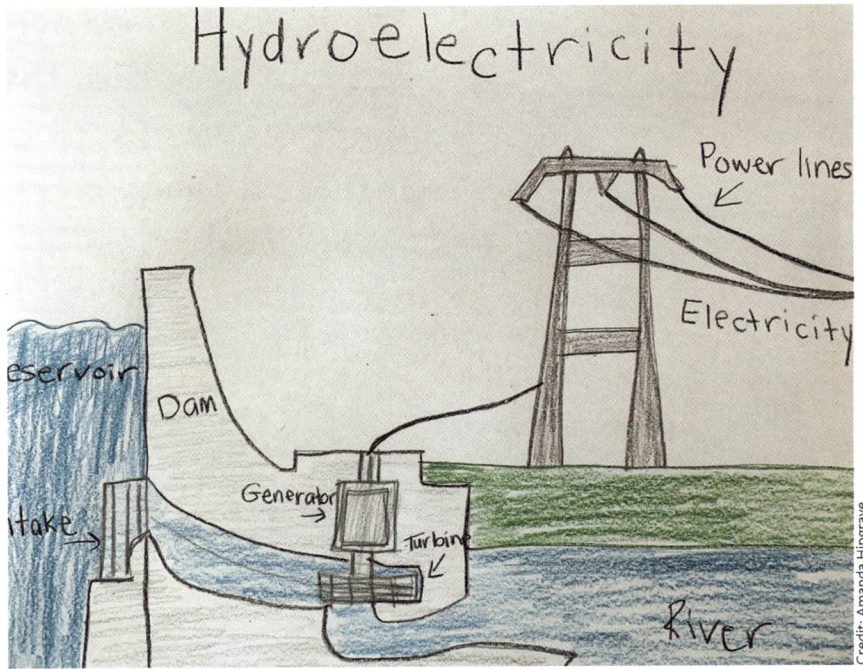

Credit: Amanda Hipgrave

● **Renewable Energy Game:** Create a board game or simulation where students make decisions about energy sources for a fictional community. The goal is to meet the community's energy needs while minimizing environmental impact. Students can learn about trade-offs and the importance of balanced energy solutions.

We participated in the local fiber production activity (video and Zoom call, exploring samples of sheep and alpaca wool) as a source of renewable clothing fibers. Students were very curious about the sheering process. They learned a lot about farming sheep and alpacas.

—SHANNON CANNON (teacher)

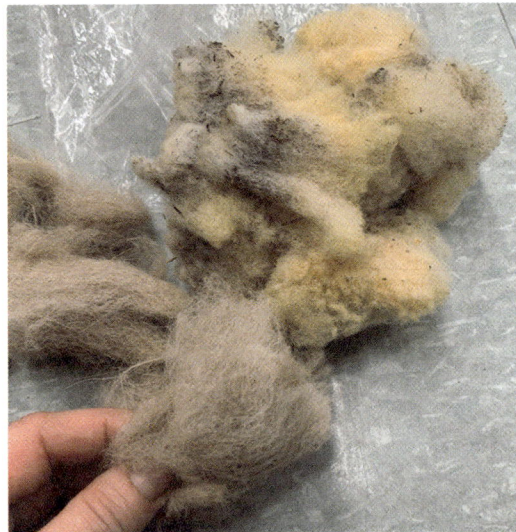

Credit: Shannon Cannon

Samples of sheep and alpaca wool

AGES 10–11 YEARS
Landmark 17

Try at least three new outdoor activities that don't require fossil fuels. Include a sport, a craft, and a survival skill.

Why?

These activities help children develop more complex outdoor skills, build confidence, and overcome fears. Learning to navigate and make use of the natural world without relying on fossil fuels fosters a sense of environmental stewardship and sustainability. These experiences strengthen personal identity and encourage a deeper connection to the land and its history. Children gain an appreciation for the traditions and knowledge of Indigenous Peoples and other cultures that have adapted to and thrived in their environments. Overall, these activities cultivate a sense of kinship with the environment and inspire a lifelong commitment to protecting and preserving our natural world.

Climate Change Connections

It has only been within the past 100 years that humanity has become so dependent on fossil fuels. For millennia, people travelled, grew food, made tools, heated their homes, and entertained themselves without needing fossil fuels. This Landmark is an opportunity to explore alternatives to fossil fuels in even more aspects of our lives—entertainment, arts, and survival.

How?

● **Try a New Sport:** Try something you've never done before: archery, snowshoeing, canoeing, lacrosse. Which of these were developed by First Peoples?

● **Fire Building:** Learn how to build a fire using only natural materials found nearby. Challenge your friends to see who can boil a pot of water first. Try your hand at making a fire using a bow drill—a traditional way of starting a fire.

● **Survival Shelter:** Create a survival shelter that will protect you from the elements. Find a forest, backyard, or park with fallen leaves and branches. Rake a large mound of dried leaves at the base of a tree with a fork in the trunk about waist high. Lean a sturdy branch (roughly 10 feet long, 5 inches wide at the base, 2 inches at the tip) against the fork and over the mound. Gather "branchy" branches to create a frame on both sides of the main branch. Pile leaves, evergreen branches, and other materials over the frame up to the thickness of your arm. Leave a pile of leaves or branches to seal the entrance once inside. This insulated hut, called a "debris hut," like a squirrel's nest, can keep you warm even in sub-zero temperatures. Thank the squirrels for the tip!

● **Stalking:** Learn the art of stalking, or walking quietly, through the woods. Here is how: Crouch slightly, hands on knees, to freeze if needed. Start with weight on the back leg. Take small steps, easing weight onto the toes of the front foot. Ensure the ground is clear of crunchy leaves or sticks. Transfer weight to the toes, roll along the foot's outside edge to the heel. Repeat with the other foot, moving slowly and deliberately. Here is a quick stalking game for a group: Two blindfolded volunteers crouch five yards apart. Participants stalk quietly between them using the stalking technique. If the volunteers hear a sound, they point to its direction. If they point at a participant, that person sits down. Who can successfully go past the volunteers?

● **Mini-Olympics:** Try organizing a mini-Olympics at your school with sports that students can organize. Senior students can plan and lead events for younger classes.

Not all classrooms have four walls.
—MARGARET McMILLAN (early learning pioneer)

● **Geocaching:** Try geocaching, using GPS to find hidden containers—can you find any geocaches in your area?
● **Traditional Northern Games:** Try participating in some northern games. These games involve a series of challenges that involve strength, agility, balance, and laughter. Many games come from the Inuit culture of Northern Canada and Alaska. Try

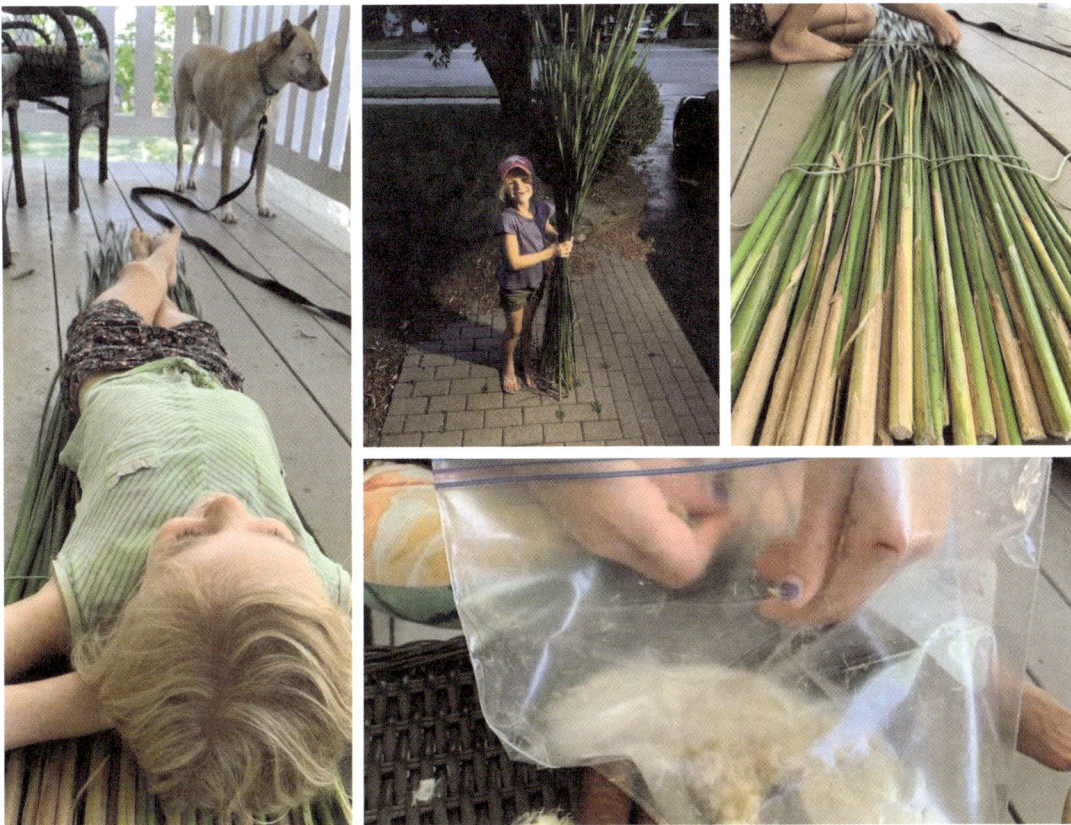

Credit: Candace Passey

Junior and Intermediate students participated in Olympic Games activities, including curling, snowshoe races/biathlon, a hockey shootout, and luging. Students and staff learned new skills and had an absolute blast! Grade 6/7 students said it was the "best school day ever."

—Amanda Hipgrave (teacher)

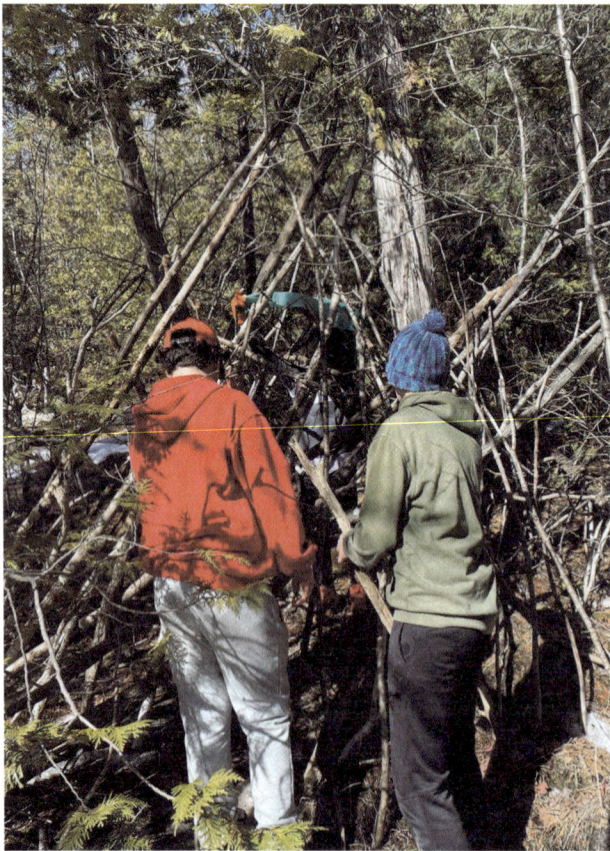

Building outdoor shelters

Credit: Mary Hollingworth

the one-foot-high kick, the knee jump, the bench reach, among many others. Here is a simple Northern game you could try right now: Try grabbing your toes. Now, without letting go, jump as far as you can. How far did you get? There are also many traditional First Nations games that are both challenging and fun. To ensure you teach and play these games respectfully, connect with a member of a First Nations or Inuit community for guidance on appropriate practices.

● **Try a New Sport**: Disc Golf: Set up a disc golf course in your local park or backyard. All you need are some flying discs and makeshift targets (like trees or poles). Disc golf is a fun way to enjoy the outdoors, and it's a sport that doesn't require any fuel!

● **Cordage Making:** Learn how to make rope or string from natural materials like plant fibers or tree bark. This ancient skill is not only practical for survival but also a fun craft to try. Use your homemade cordage to create a bracelet, keychain, or even a bowstring.

● **Outdoor Cooking:** Learn to cook over an open fire without modern conveniences. Try baking bread on a stick, roasting vegetables, or even making a simple stew. Cooking outdoors teaches you how to use natural resources efficiently and is a delicious way to enjoy the fruits of your labor.

● **Nature Crafts:** Try making a craft using only natural materials: try stringing a weaving frame between a square of sticks, and weave natural materials to make a beautiful wall-hanging. Is there someone in your community who can help you learn how

Our 5/6 classroom has been fortunate to host a teacher candidate who has shared a variety of Inuit games with us. Through these activities, students have been challenged to develop their skills related to agility, strength, resilience, and perseverance, all while connecting to and deepening their respect for the land.

—KARA YORK (teacher)

We brought the tall reeds home, and with string I showed my mom how to tie them into a cattail mat. First Nation Peoples used this technique to make beds, baskets, and clothes. I also fluffed the brown cattails and put it in a bag. We brought it camping and used it as a fire starter.

—ISLA PASSEY (student)

to make a simple basket using grasses and reeds? Or, try knitting a simple square or scarf using natural yarn. What other crafts can you think of that use natural materials?

● **Pinch Pot:** Try making a simple pinch pot out of clay. Use store-bought air dry clay or use natural clay if you can find some near where you live. Work the clay in your hands, rolling and pounding until it be- comes easier to shape. Form a ball that fits easily in your palm. Stick your thumb in the middle of the ball of clay and press down until it is about halfway. Using your thumb and fingers, pinch and shape until the walls of your pot become thinner and taller. If your pot begins to crack, add a bit of water to your fingers and continue working the pot until you have the desired shape. Use shells, seeds, or the tips of a stick to gently press into the outside of your pot to give it an interesting texture. Allow to air dry.

Credit: Geri-Lyn Cajindos

AGES 10–11 YEARS
Landmark 18

Celebrate a local natural area. Create a book, blog, or video about a nearby natural area to encourage people to visit and appreciate it.

Why?

Thinking about what makes a natural area unique and beautiful sparks deeper discussions about our values and our relationship with the natural world. By observing and reflecting on these spaces, students can engage in meaningful conversations that help them articulate what they find significant and why it matters to them. This kind of exploration is an excellent class project that invites diverse opinions and encourages students to express their perspectives in creative ways, such as through art, writing, presentations, or multimedia projects.

At this stage of development, young people are beginning to explore what shapes their values and their relationships with other living things. They can examine questions like: What do we value in nature? Why is it important to protect natural spaces? How do our actions reflect our values? Through these inquiries, students can develop a deeper understanding of their role in the ecosystem and the significance of fostering a respectful and caring attitude toward the natural world. This process not only enhances their appreciation for nature but also cultivates a sense of empathy and responsibility that can influence their behavior and decisions in the future.

Climate Change Connections

Natural areas are important for many reasons, including their ability to take carbon out of the air and store it in plant and animal tissues. Natural ecosystems can extract as much as one third of the annual carbon dioxide added to the atmosphere through human activities. That makes natural areas critically important as climate change mitigators. As human populations continue to grow, there is increasing pressure to build on natural areas or extract their contents for economic gain. In a culture that often equates "value" with "economic worth, it's important to remember the many other things in life that have value in different ways. As a further

A weed is no more than a flower in disguise.

—James Russell Lowell (poet)

extension of earlier Landmarks that develop a connection to place, Landmark 18 encourages finding ways to share a love of a natural area so that others can find value in it too.

How?

● **Nature Travel Guide:** Make an inviting travel guide for a nearby natural area that you like to visit; include photos and written descriptions of why it's a great destination. Create ideas on how your natural area could be used during different seasons. Invite a local scientist or ecologist to share their knowledge of the area that you might use in your travel guide.

I am a…tree hugging, flower sniffing, animal kissing, planet loving, dirt worshipper and proud of it!

—Anonymous

● **History of the Land:** Research the history of this place: who lived here before and how did they care for the land? Use this as an opportunity to learn more about the Indigenous Peoples in your area. What can we learn from First Nations about how to have respectful and reciprocal relationship with the natural world?

● **Share Your Inspiration:** Work with others to produce maps, artwork, stories,

Credit: Karen Brown

or poems for your book or blog to reflect what you learned and how you feel about the area. Did you find that everyone sees the same things as beautiful or interesting? If not, what do you think contributes to these differing opinions?

● **Place as Teacher:** What did you learn from that spot that didn't come from a book or a screen? Perhaps you observed how a spider weaves its web or how a beetle moves across the forest floor. There is something powerful that comes from learning directly from the natural world itself.

● **Tending:** Discuss how you could encourage others to visit, enjoy, and protect the area. Is there something you can do as a class to help this area (removing invasives, planting native species, creating a local field guide, teaching younger students).

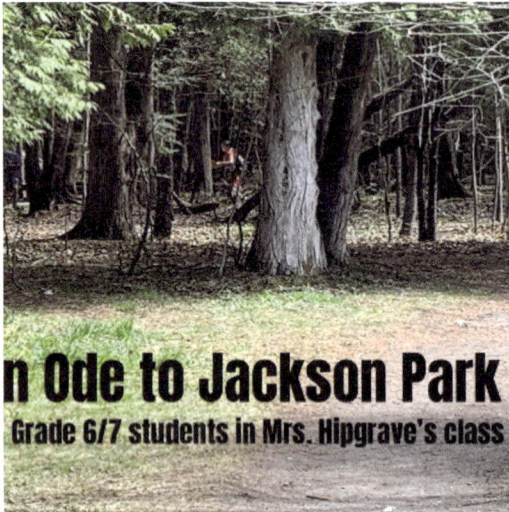

n Ode to Jackson Park
Grade 6/7 students in Mrs. Hipgrave's class

Nature
Lovely woodpecker
Beautiful, big, bushy, bird
Peaking holes in wood

Water looks so clear
it's blue like the sky my dear
and I love it here

Fly away birdy
Are you here in danger bird
Okay fly away

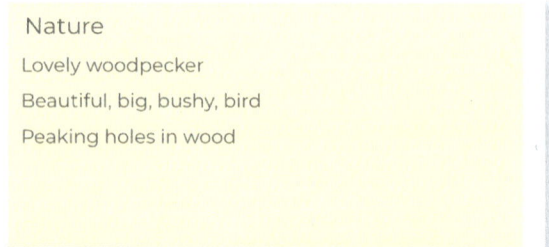

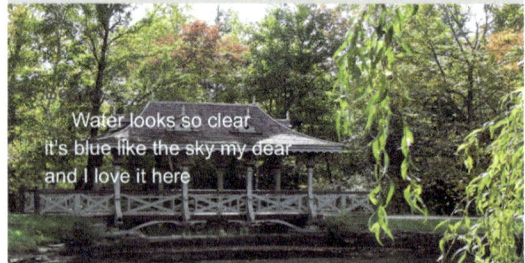

The grass can grow high
The grass attracts worms and bugs
They make soil for us

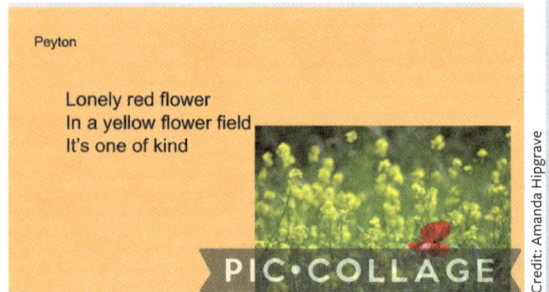

Oh, cold moving wind
Making the trees dance around
With such happiness

Peyton

Lonely red flower
In a yellow flower field
It's one of kind

PIC·COLLAGE

Credit: Amanda Hipgrave

- **Nature Scavenger Hunt:** Design a scavenger hunt with a list of items, plants, animals, and natural features for visitors to find. This can make exploring the area more interactive and fun for families and children.
- **Photography Contest:** Encourage visitors to capture the beauty of the natural area through photography. Organize a contest and showcase the best photos in a local exhibition or online gallery.
- **Guided Nature Walk:** Plan a guided tour of the natural area led by knowledgeable volunteers or local experts. These tours can highlight the unique features and biodiversity of the area.
- **Seasonal Events and Workshops:** Organize events and workshops throughout the year, such as bird-watching in the spring, mushroom foraging in the fall, or stargazing nights in the summer. These activities can highlight the seasonal changes and diverse experiences the area offers.
- **Local Artists:** Invite local artists to create works inspired by the natural area. Host an art show or create a mural that captures the essence of the place.
- **Children's Activity Book:** Create an ac-

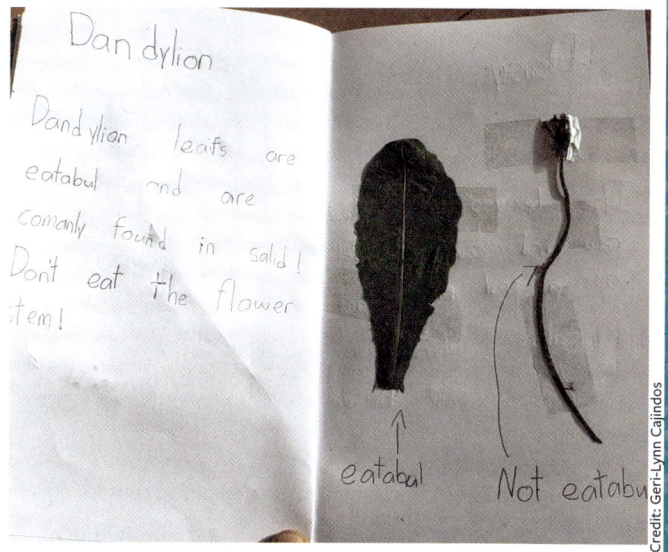

Credit: Geri-Lynn Cajindos

tivity book for children with coloring pages, puzzles, and games related to the natural area. This can make learning about nature fun and engaging for young visitors.

- **Time-Lapse Video:** Document the natural area over a year with a time-lapse video, showing the changing seasons and highlighting the area's dynamic nature.
- **Storytelling Sessions:** Invite local storytellers, elders, or cultural leaders to share stories and myths about the natural area, enriching visitors' understanding and appreciation of its cultural significance.

For National Poetry Month in April, students wrote and compiled a book of poetry to celebrate Jackson Park. They were inspired by a nature walk and wanted to capture the moments that stood out to them the most.

—AMANDA HIPGRAVE (teacher)

AGES 10–11 YEARS
Landmark 19

Explore biodiversity. Find out what lives in a nearby wetland, forest, or meadow.

Why?

Understanding biodiversity fosters a sense of wonder and respect for the complexity of life. When children investigate which plants provide food, shelter, or nesting sites for insects, birds, or mammals, they begin to grasp the concept of ecosystems—how everything is interconnected. They can learn how the disappearance of one species, such as a specific pollinator, can ripple through the environment, affecting plants, animals, and even humans.

Learning to identify species sharpens children's observation skills by encouraging them to pay attention to detail. As they distinguish subtle differences between similar organisms—such as variations in leaf shapes, bird calls, or insect patterns—they develop a keener eye for the intricacies of nature. This process not only enhances their ability to notice and appreciate the richness and diversity of natural systems, but it also deepens their understanding of how each species plays a unique role within its ecosystem.

Climate Change Connections

The Earth is undergoing a dramatic loss in biodiversity as a direct result of human activity. In just 50 years, we have lost a staggering 70 percent of the Earth's wildlife. Many of those that remain are facing enormous pressures in trying to adapt to increased changes in climate and loss of habitat. Young people deserve many opportunities to continue to experience a sense of awe and wonder for the amazing diversity and beauty of life on Earth. This can be the fuel that drives a desire to help seek solutions to environmental and social challenges. Any exposure to environmental problems must always be balanced by presenting options for solutions. Even though loss of biodiversity is one of the biggest challenges on Earth today, problems caused by people must also be solved by people. Young people need to know that there is hope, and that they can and must be part of a change in how we interact with the world that supports us. That can begin in our own homes and communities.

Credit: Amanda Higgrave

How?

● **Local Frogs:** Get to know your local frogs. Frogs can be found on every continent except Antarctica and live wherever there is fresh water. Frogs are also bioindicators. Because they breathe partially through their skin, they are sensitive to chemical pollutants in the water. So, if a pond is contaminated, there won't be many frogs. On the other hand, a healthy pond will support a healthy frog population. You can learn to identify frogs by the sound they make. Each species of frog has a unique call. Visit a wetland in early spring and listen for frog songs. You'll find they call more in evening. How many different kinds of calls can you hear? There are many online resources that will help you identify which frogs are calling. Report your findings to a Citizen Science website such as FrogWatch. Go back a month later and see if you hear different species of frogs singing.

● **Transect Walk:** Select a natural area near your school. Have each student walk a transect or a line through the area and identify

> **You should sit in nature for 20 minutes a day...unless you are busy, then you should sit for an hour.**
>
> —Zen saying

> **If we want children to flourish, to become truly empowered, let us allow them to love the Earth before we ask them to save it.**
>
> —DAVID SOBEL (educator and author)

what you see around you every 5 paces. Each transect should be about 10 paces apart. Record your findings. Now use the data you've collected to create a map of biodiversity. Is your natural area nature rich? Is there a way that you might add to the biodiversity by planting more native species?

● **Nature Buddies:** Explore a local forest or meadow by matching older students with their younger "buddies." Have each pair look for as many different living things as they can. Keep a record of what was found.

● **Benthic Invertebrates:** Visit a local stream with dip nets, buckets, and magnifying glasses. Look for living things in the sediments; collect them in your net and place gently in a bucket of clean water. How would you describe the animals you found? Conduct some research to find out more about what they eat and how they are adapted to living in streams. Often your local conservation organization has field guides to help identify your specimens. The presence of certain aquatic invertebrates can indicate water quality. Based on the critters you've caught, what does that tell you about the health of your stream? Be

Credit: Lisa Fitzsimmons

sure to release the animals back to where you found them.

● **Tree Characters:** How many different kinds of trees can you find in a nearby forest? Compare the bark, branches, and leaves. Create a character and story for each tree species. For example, you might name a white pine "Five Finger Freddy" because its needles grow in clumps of five. You can make a refreshing tea rich in vitamin C with the needles. White pine resin has antibacterial properties, and the sound of the wind through the boughs is a musical and gentle "whoosh!" What names and stories can you come up with for your local tree species? Do different trees grow in wet places than dry places? Do some species of trees prefer sun or shade?

● **Moth Light:** When night falls during midsummer, hang up a white sheet under a strong light. Wait and it won't be long until a variety of moths land on your sheet, attracted to the light. Use your Seek app to see if you can identify any of these and post to iNaturalist, so that scientists can use the data. Marvel at all the different sizes, shapes, and colors of moths that come to visit!

● **BioBlitz:** Organize a BioBlitz at your school. Have each class work together to identify everything thing that lives in and around your school. The apps Seek or iNaturalist can be a great help in identifying what is living around you. A local

On summer solstice, we slowly explored our yard on hands and knees, peeking at the various plants and the animals on the plants. We found lots of interesting things on our milkweed plants in particular (beetles, aphids, weevils, thrips, native bees, flies, and a spider)!

—LISA FITZSIMMONS (parent)

naturalist from your school community or naturalists' club can also be a wonderful resource. Create a map and update this on an annual basis. Over time, are you able to help make your school more biodiverse?

Credit: Candace Passey

Landmarks for Age 12 to 13 Years

Age 12 to 13 Years: Characteristics of This Age Group

Social interactions take center stage at this age. Typically, these young teens are in the early stages of puberty, making them very self-conscious and afraid of making mistakes. This self-awareness can sometimes make them less responsive in a classroom setting or at home, particularly if they fear judgment from peers or family. However,

Credit: Invato: mstandret

their strong affiliation with peers means they are highly influenced by what their peer group thinks, believes, and feels.

At the same time, 12- to 13-year-olds are searching for their personal identities as they experience hormonal shifts and a growing need for independence. This quest for identity can make them more open to exploring new ideas and causes, especially those that align with their emerging values and desire to make a difference. They are beginning to form deeper ethical and moral perspectives and refining their core beliefs.

Young people at this age enjoy problem-solving and often work well as a team, particularly when they are given ownership of a project or challenge. They also can be excellent mentors to younger children, taking on roles such as reading partners or games buddies. This age group also benefits from opportunities to play a leadership role in environmental education. Mentoring younger students on environmental topics not only reinforces their own learning but also helps them develop a sense of responsibility and pride in being role models.

© linda_vostrovska / Adobe Stock

AGES 12–13 YEARS

Landmark 20

Plan, conduct, and evaluate at least two environmental projects.

Why?

To further develop motivation and empowerment, young people need opportunities to be leaders and decision-makers in their communities. Engaging in local action with projects that are small in scope and easily accomplished encourage our youth to feel a sense of accomplishment and help them to recognize that they can be powerful agents of change. Planning, leading, and reflecting on projects are wonderful opportunities to learn, grow, and build a sense of hope. Hope, along with tangible action, is the first step in creating meaningful change.

Climate Change Connections

The action ideas listed below are just a place to begin. Do you have other ideas for a project you could do with your family or class to help address climate change? Remember that writing or speaking to local politicians or other community officials is important in expressing your views on this issue that effects all of us.

How?

● **Schoolyard Mapping:** Complete a map of your schoolyard and collaborate with your parent council, administrators, and fellow students to develop a long-term plan for creating spaces where both wildlife and people can thrive. Aim to add a bit each year, gradually increasing biodiversity over time. Think about incorporating rewilded areas, pollinator gardens, sensory gardens, an arboretum of local tree species, and bird-friendly areas with plants that attract birds, along with bird feeders. Consider creating special spots for people to reflect and enjoy nature, an outdoor classroom, and a small nature trail.

● **Nature Day Trip:** Plan and map a local day trip to a natural area for your class or family; use sustainable forms of transportation to get there (walk, bike, canoe, bus, etc.). Bring a picnic on your trip and reflect on what you experienced when you return home or school. Encourage Walk to School days. Calculate how much carbon a class uses collectively as they travel to school over the school year. Challenge each other to reduce this footprint.

● **Plastic, Plastic:** When next you shop, count how many objects are encased in plastic. What are the impacts of so much plastic? Does some of it get recycled? What happens to the rest? Do you have some ideas on how to reduce the amount of plastic in the store? Share these ideas with the owner of the store, with environmental groups, and with your local politicians.

● **Product Life Cycle:** Research the life cycle and environmental impacts of an item or fashion product you use regularly. How much waste and energy does it take to create this product? What will happen to it at the end of its useful life? What are some ideas you might have to extend the life of this product, to reuse or repurpose it, or to make it more sustainable? Write a letter to the company that makes this product and share these ideas.

● **Waste Audit:** Do a waste audit of your classroom. How much material ends

One individual cannot possibly make a difference, alone. It is individual efforts, collectively, that makes a noticeable difference—all the difference in the world!

—Dr. Jane Goodall (zoologist and primatologist)

The best education does not happen at a desk, but rather engaged in everyday living—hands on, exploring, in active relationship with life.

—Vince Gowmon (author)

Credit: Lindsay Bowen

Intermediate students leading school fun day

Our class is taking part in the Lake Ontario Atlantic Salmon Restoration Program-Classroom Hatchery Program. This program is delivered to us virtually through the OFAH this year. We are learning about the life cycle of the salmon and watching the whole process through video conference. We will hopefully have our own aquarium with eggs next year, to have a chance to take care of the fry, watch them grow, and help release them into Lake Ontario.

—HEATHER MORTON (teacher)

Our intermediate students planned and ran outdoor activities for the whole school for a whole week! This project lifted the spirits of every student and staff and got everyone outside enjoying the brisk spring air.

—LINDSAY BOWEN (teacher)

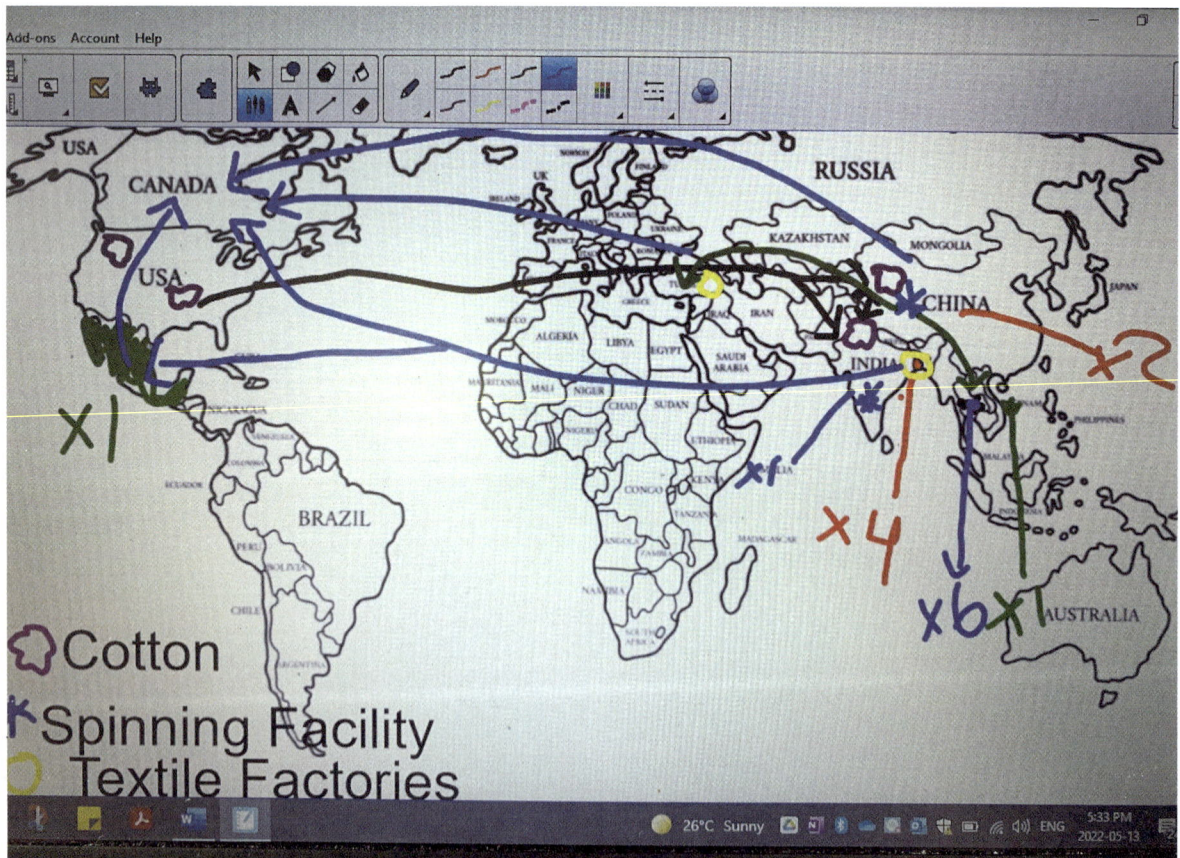

Researching cotton transport

The most environmentally friendly product is the one you didn't buy.

—Joshua Becker (author)

up in the garbage? Can more waste be recycled? Think about setting up a small vermicomposter—to help you turn your food scraps into soil.

● **Conservation Project:** Participate in a local conservation project. This can include monitoring wildlife populations, planting trees, and building a garden for wildlife. Share these stories with the local press and with your school administration.

● **Idling Campaign:** Start a campaign to reduce vehicle idling around your schoolyard. When vehicles are left running while they're not moving, a great deal of fossil

Students completed the Schoolyard Report Card. They drew a map, filled in a Report Card ("Habitat for People" and "A Healthy Environment"), and are making recommendations to our school principal.

—Karen Brown (teacher)

fuels are wasted, which also adds to the problem of climate change.

● **Family Energy Challenge:** Work with your family to see if you can reduce your energy consumption for one month. Talk together about some of the things you could try. Set up a chart where you can keep track of each effort you make. After one month, talk together about how easy or difficult it was, and which things you could continue doing on a regular basis.

Credit: Karen Brown

Researching manufacture of fashion products

AGES 12–13 YEARS

Landmark 21

Learn about at least two other cultures by meeting and talking with someone whose culture is different from yours.

Why?

Diversity matters. That is why it is important to provide opportunities throughout childhood to develop meaningful and respectful relationships with people from diverse backgrounds and cultures. This becomes especially important in the early teen years when peer groups are the focal point of social interaction. In our interconnected world, peace and harmony depend on understanding that there is no "us" and "them." We are all inextricably linked, and learning to value human diversity is as crucial as valuing broader biodiversity. Encouraging young people to engage with individuals from different cultures fosters empathy, broadens perspectives, and cultivates a sense of global citizenship. By meeting and talking with someone whose culture differs from their own, youth can gain a deeper appreciation of the rich tapestry of human experience, reinforcing the idea that diversity is a strength to be celebrated and respected.

Climate Change Connections

Truly addressing the challenge of climate change requires a great deal of international cooperation. The causes of climate change come from all corners of the world, and likewise, action must be taken globally.

We can only work effectively together if we bring a true sense of respect for each other to the discussion table. Concern for the health of our children, today and tomorrow, is something that unites us all. All young people benefit from many opportunities to make friends with people from many cultures.

How?

● **Elders:** Invite local Indigenous Elders and knowledge holders to play a role in the education of all children and youth. Encourage them to tell you the story of their people and how settlement impacted their community and culture. Traditional ways of knowing have much to teach us about how to have a reciprocal and respectful relationship with each other and the Earth we share. What are some teachings that you can implement in your classroom or at home right now?

The things that we share in our world are far more valuable that those which divide us.

—DONALD WILLIAMS
(former U.S. astronaut)

Our school enjoys the presence of many New Canadian friends. Several of them are currently observing Ramadan. Two students from our class are doing daily announcements. We are learning new things every day. Thank you Tahani and Gulbinda!

—Karen Brown (teacher)

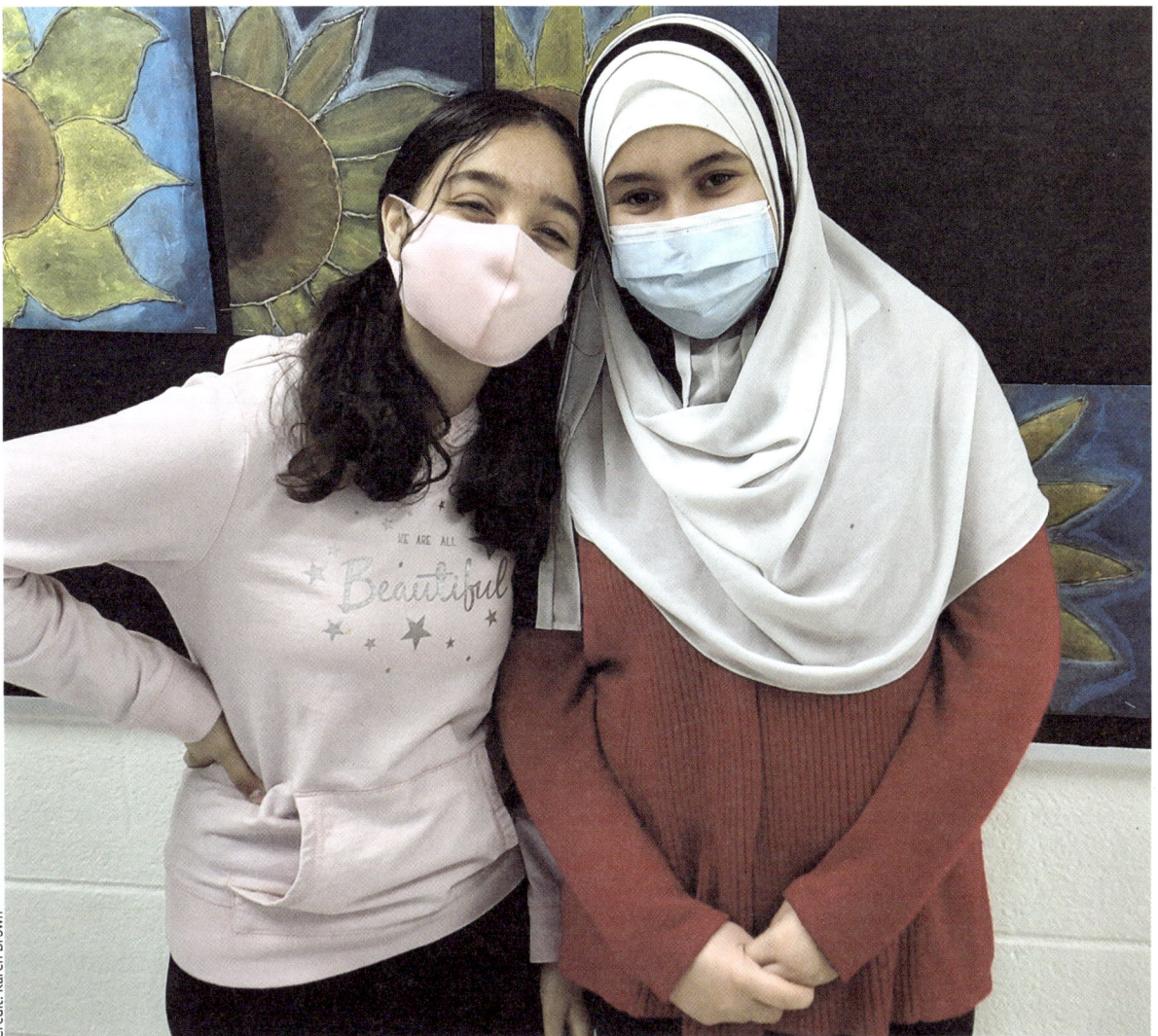

Credit: Karen Brown

Students sharing daily reports on the customs of Ramadan

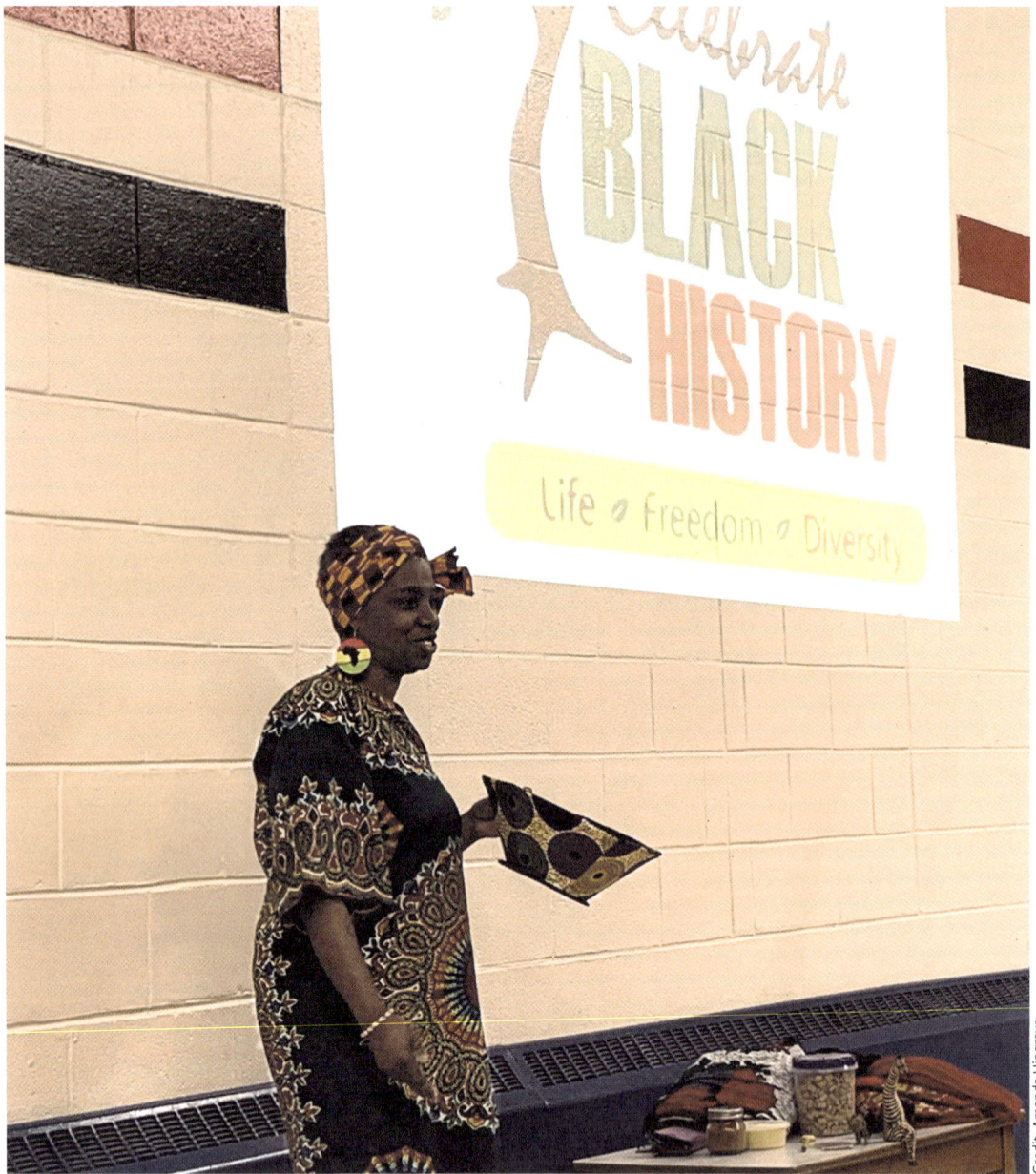

Credit: Amanda Hipgrave

Intermediate students attended an assembly to learn about Black History Month, the African diaspora, Jamaican culture, and Viola Desmond. A big thank you to Charmaine Magumbe for her time, knowledge, and kindness!

—AMANDA HIPGRAVE (teacher)

• **Indigenous Events:** Participate in special events held in local Indigenous communities such as making maple syrup, harvesting wild rice, or attending powwow celebrations.

• **Traditions:** Encourage people in your community to share the foods and traditions of their homeland.

• **Stories of Immigrants:** Invite immigrants to share their experiences of leaving their homeland and making a home in a new place.

• **Seniors' Home:** Visit a seniors' home and write the story of one senior's life. Compile the stories in a self-published book and share with the seniors' home.

• **Special Events:** Plan special events to honor annual recognition days such as National Indigenous Peoples' Day in June or Black History Month in February.

• **Cultural Nature Connection:** Explore the kinds of relationships newcomers to your country have had with nature where they live. What are some similarities and differences between how they view the natural world and how we treat nature? What can we learn from them?

• **Cultural Exchange Programs:** Set up exchange programs where students spend time with families from different cultures to experience their daily life, traditions, and customs firsthand.

• **Multicultural Festivals:** Organize school or community festivals that celebrate various cultures through music, dance, art, food, and traditional crafts. Invite community members to participate and share their heritage.

> **When I see you through my eyes, I think that we are different. When I see you through my heart, I know we are the same.**
>
> —DOE ZANTAMATA (author)

• **Cultural Storytelling Sessions:** Arrange sessions where people from different backgrounds share folktales, myths, and stories from their culture, fostering an understanding of their history and values.

• **Community Projects:** Encourage joint projects between different cultural groups, such as community gardens, murals, or environmental cleanups, to build relationships and work toward common goals.

• **Cultural Pen Pals:** Establish pen pal programs with schools or communities in different countries to encourage cultural exchange and understanding through regular communication.

• **Film Screenings and Discussions:** Host screenings of films and documentaries that highlight different cultures and their stories, followed by discussions to deepen understanding and empathy.

• **Collaborative Art Projects:** Initiate collaborative art projects where participants from diverse backgrounds create art together, reflecting their cultures and experiences, and promoting unity through creativity.

• **Mentorship Programs:** Pair students or community members with mentors from different cultural backgrounds to build relationships and share knowledge and experiences.

AGES 12–13 YEARS

Landmark 22

Become a Citizen Scientist. Participate in a program to collect and submit data about the local environment.

Why?

All around the world, scientists are making amazing discoveries, and people are helping them through the application of Citizen Science. Citizen Science involves everyday citizens collecting data that scientists use in their research. Engaging in Citizen Science provides youth with an ideal opportunity to combine outdoor observation with the math skills required for data collection and analysis. This hands-on experience fosters a developing sense of agency in students, empowering them to understand that they can influence the world around them. At the same time, it helps them see real-world applications for the math skills they are learning in school. There are many programs for students that encourage citizen participation throughout all seasons, guiding a wide variety of conservation initiatives and making significant contributions to environmental stewardship.

Climate Change Connections

This is another opportunity to build young people's confidence to participate in local environmental action. Too many young people feel helpless in addressing big environmental issues such as climate change, and need many opportunities to start with small manageable real-life tasks to realize that everyone can help in some way, and it takes many people getting involved to bring about meaningful change.

How?

- **Help with Nature ID:** Try user-friendly programs such as iNaturalist to help identify things in nature and submit information about local ecosystems.
- **Monarch Watch:** Classes can assist with monitoring monarch butterfly populations by raising monarchs and reporting sightings to Monarch Watch.
- **Migration Watch:** Students can track the migration of birds, butterflies, and even whales by visiting Journey North.
- **Amphibian Watch:** In spring and summer, students can participate in monitoring local amphibian populations through Frog Watch. Frogs are considered "bioindicators," meaning their presence typically signals good water quality. By identifying and counting different frog species in nearby wetlands, students contribute to assessing the health and biodiversity of these vital ecosystems.
- **Water Quality Monitoring:** Local conservation organizations can teach students basic techniques for monitoring water

quality. Students can learn to monitor dissolved oxygen, pH, salinity, transparency, temperature, and conductivity and use these to help determine the health of the local watershed.

● **Tree ID:** Students can work with local experts to learn tree identification skills, and then help to submit tree inventories that assist with urban forest planning.

● **Weather Monitoring:** Students can participate in weather tracking projects like CoCoRaHS (Community Collaborative Rain, Hail, & Snow Network), where they measure and report precipitation.

● **Pollinator Counts:** Engage students in monitoring pollinator activity, such as bees

> You are not Atlas carrying the world on your shoulder. It is good to remember that the planet is carrying you.
>
> —VANDANA SHIVA (environmental activist)

> We raised, tagged, and released 8 monarch butterflies. We also submitted a spreadsheet of our data to Monarch Watch.
>
> —SHEILA POTTER (teacher)

Credit: Sheila Potter

Education is not the filling of a pail, but the lighting of a fire.
—W.B. Yeats (poet)

and other insects, through projects like the Great Sunflower Project.

● **Phenology Tracking:** Students can record seasonal changes in plants and animals, contributing to projects like Nature's Notebook, which tracks phenological events such as blooming and migration.

Testing local water quality

Credit: Janet Gray

● **Marine Life Surveys:** If located near coastal areas, students can participate in beach cleanups and marine life surveys, contributing data to organizations like the Ocean Conservancy.

● **Urban Wildlife Monitoring:** In urban settings, students can document and report sightings of local wildlife through platforms like Project Squirrel or the Urban Wildlife Information Network.

● **Air Quality Monitoring:** Collaborate with environmental organizations to set up air quality monitoring stations and have students collect and analyze data on air pollutants.

● **Soil Health Projects:** Students can learn about soil health by collecting samples and analyzing them for various properties, contributing to projects like the Global Soil Biodiversity Initiative.

● **Light Pollution Tracking:** Students can participate in projects like Globe at Night, where they measure and report light pollution in their area to help scientists understand its impact on ecosystems.

● **Invasive Species Monitoring:** Engage students in identifying and reporting invasive plant and animal species through platforms like the National Invasive Species Information Center.

● **Citizen Science Festivals:** Organize or participate in local citizen science festivals where students can engage in various projects, learn from experts, and share their findings with the community.

Credit: Leanne Kelly

● **Birds, Birds, Birds:** eBird is a wonderful program to track bird populations around the world. It's simple to report the birds you've seen on any given day, and the reports are used to monitor trends in bird activity and to plan suitable conservation programs. There's a corresponding app called Merlin that can help you identify any bird you see or hear, and it's equally helpful for beginners and experts alike.

We learned about the Otonabee Region Watershed, indicators of watershed health and measured chemical and physical properties of water.
—Janet Gray (teacher)

© RM Graphics / Adobe Stock

AGES 12–13 YEARS
Landmark 23

Design your own healthy house.

Why?

Using mushrooms to create ceiling tiles? Making floors out of eggshells, window-sills out of recycled textiles? There are so many creative ways to use natural materials to create healthy and sustainable homes. This is a chance for students to do some research on bio-based buildings, on energy-efficient design, and on alternative forms of energy. And, to engage their creativity to design something truly innovative. As young people continue to look at human relationships with the natural world, it makes sense to include our own habitats in the discussion. How can we live in ways that are kind to the environment, including how we build and maintain our own homes, both inside and out? This is an ideal research project: to explore sustainable building options to create living spaces where both people and nature can thrive.

Climate Change Connections

There are many exciting developments taking place in sustainable building design and construction around the world. Conventional building design adds to the burden of greenhouse gases through extensive use of concrete, felling of trees for wood, and pro-duction of a wide range of insulation and finishing chemicals that are not only toxic but require fossil fuels for production. Many conventional buildings also require considerable fossil fuels or electricity for heating and cooling. Sustainable building materials such as straw bale and hempcrete are climate friendly, and buildings can be designed to use solar or geothermal energy for heat or cooling to dramatically reduce nonrenewable energy consumption. This Landmark is an excellent opportunity for young people to explore these new technologies while exercising their own creative design skills.

How?

● **Healthy Versus Unhealthy Homes:** Discuss what makes a house healthy or unhealthy. How can an existing house be

A vibrant, fair, and regenerative future is possible—not when thousands of people do climate justice activism perfectly but when millions of people do the best they can.

—Xiye Bastida (climate activist)

adapted to be healthier for people and the environment?

● **Visit a Local Sustainable Home:** What unique features does this home have? How does it differ from a conventional home? Could you live in a home like this?

● **Sustainable Building Investigation:** Research a variety of approaches to sustainable buildings; consider energy source and use, water source and use, building materials, landscaping options etc. Here are some things to consider:

- **Use Recycled and Upcycled Materials:** Research and incorporate recycled or upcycled materials into the home design. Discuss the benefits of using these materials and where they can be sourced. Research bio-based buildings for ideas on what nature-based materials can be incorporated into your design.

- **Incorporate Renewable Energy Sources:** Design systems that use solar, wind, or geothermal energy. Create simple models to demonstrate how these energy sources can power a home.

- **Water Conservation Techniques:** Plan for rainwater harvesting systems, gray water recycling, and water-efficient fixtures.

- **Indoor Plants:** Learn about plants that can improve indoor air quality and include them in the home design. Discuss the importance of ventilation and nontoxic building materials.

- **Insulation and Passive Heating/ Cooling:** Explore different insulation

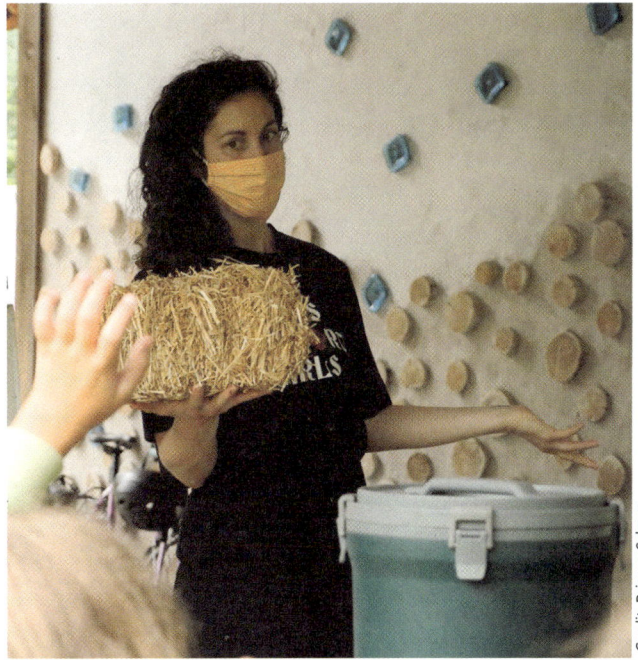

Credit: Brianna Salmon

Learning about sustainable building materials

materials and methods. Design features that maximize passive heating and cooling, such as strategic window placement and living roofs.

- **Growing Food:** Consider incorporating spaces for growing food within the home or community. Discuss the benefits of local food production and how it can reduce a home's environmental footprint.

- **EV Stations:** Plan for electric vehicle charging stations and bike storage.

Imagination is more important than knowledge. Knowledge is limited. Imagination encircles the world.

—ALBERT EINSTEIN (physicist)

Credit: Amanda Hipgrave

Dreaming of sustainable communities

Students researched and built models of an energy efficient home that would require less electricity/ energy from traditional methods. They included lots of solar panels, wind turbines, gardens, rain barrels, and energy efficient appliances. A couple really thought outside the box, including things like a library so people would spend less time on tech!

—Shannon Cannon (teacher)

GreenUP's Girl's Climate Leadership Camp participated in a workshop led by the Endeavour Centre's Jen Feigin. This workshop focused on healthy and sustainable building techniques and materials. This workshop also highlighted the important role that girls and women can have in industries such as construction, where gender-based discrimination is an ongoing barrier.

—BRIANNA SALMON
(environmental activist)

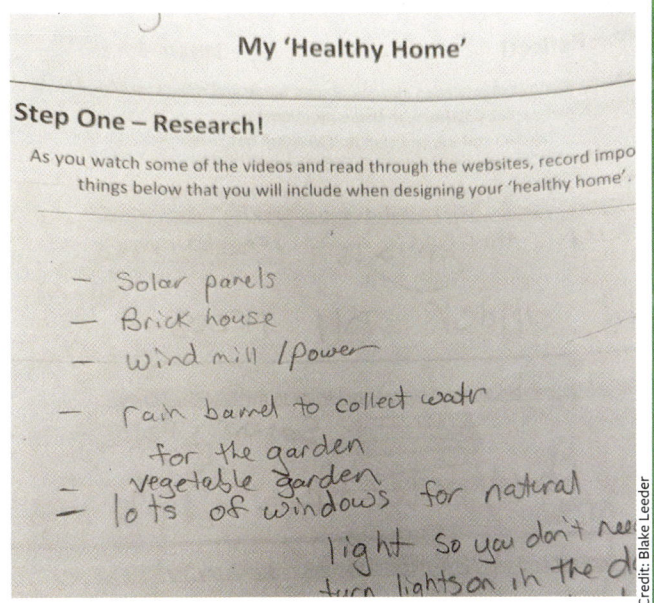

My 'Healthy Home'

Step One – Research!

As you watch some of the videos and read through the websites, record impo[rtant] things below that you will include when designing your 'healthy home'.

- Solar panels
- Brick house
- Wind mill / power
- rain barrel to collect water for the garden
- vegetable garden
- lots of windows for natural light so you don't nee[d] turn lights on in the d[ay]

Credit: Blake Leeder

Discuss how these features can reduce the carbon footprint of transportation.

- **Reducing Waste:** Design systems for composting, recycling, and reducing waste within the home.
- **Natural Light:** Think about using natural light and include elements of biophilic design (or design with love of nature in mind), such as natural materials, views of nature, and indoor gardens, to enhance well-being.
- **Living Buildings:** Explore the concept of living buildings, the design which is inspired by how a tree or a flower works. Living buildings generate more energy than they use, incorporate materials that are biodegradable, harvest rainwater without depleting lakes or aquifers, and they are designed to look as if they belong on the landscape, just a like a tree. Living buildings enhance the natural environment by increasing the presence of nature in and around the building site. Can you design a living building?
- **Sustainability and Cultures:** Explore how different cultures approach sustainable living. Think about how you might incorporate traditional sustainable practices into modern designs.
- **Sustainable Building Plan:** Draw a floor plan of your sustainable dream home and a sketch of what it looks like from the outside. Present your design to the community for feedback. This can include family members, local architects, and sustainability experts. What suggestions might they have to improve or refine your design? Combine other students' design work to create a healthy community.

Landmarks for Age 14 to 15 Years

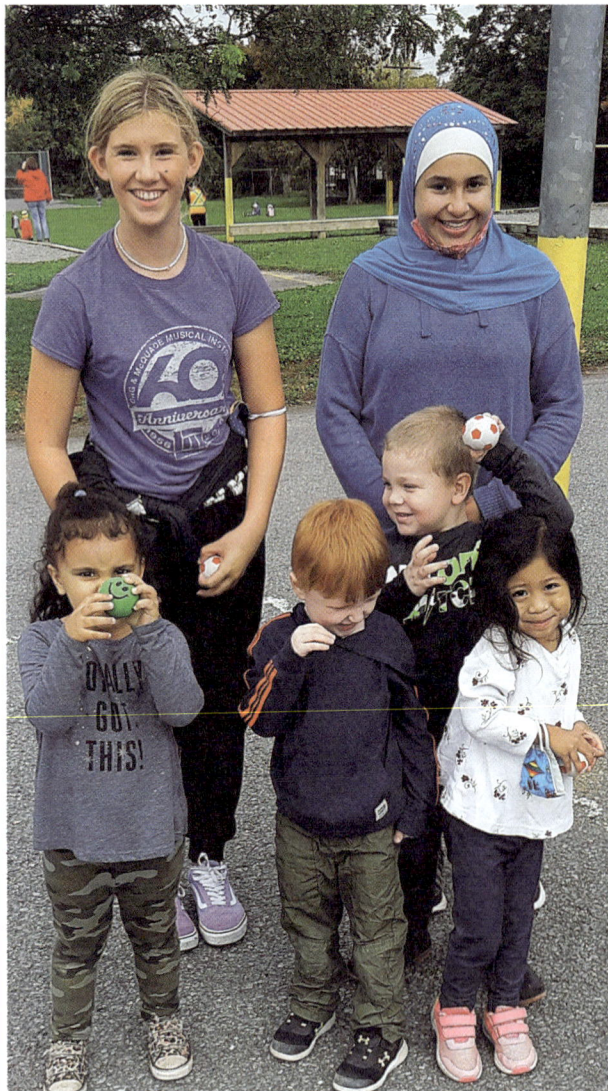

Credit: Amanda Hipgrave

Age 14 to 15 Years: Characteristics of This Age Group

This is a tumultuous time of life as young people search for their own identity, strive to belong, while navigating social pressures and social norms. At 14 to 15 years old, teens are often just beginning secondary school. After being the oldest in elementary school, they suddenly find themselves among the youngest again, which can be disorienting. Social awkwardness is common, yet they feel most at ease when spending time with close friends.

This age is characterized by rapid physical, mental, emotional, and cognitive growth. Teens at this stage yearn to belong to a group, and they often are driven by the need to fit in and be accepted. The intensity of their emotions increases, with many experiencing strong sexual attractions and a budding interest in romance.

Youth at this age are naturally inclined to test boundaries and seek out new adventures, eager to explore the limits of their abilities and independence. As their cognitive abilities mature, they begin to understand the complexities of the world around them, leading to a deeper desire to find their place within it. Young teens continue to develop their mentorship skills. As they explore leadership roles, they start to understand what it means to be responsible and to care for others. This sense of responsibility, combined with their expanding worldview, can inspire them to take on challenges and actively contribute to their communities, making this an ideal time to engage them in environmental stewardship and social justice issues.

© Olha / Adobe Stock

136

AGES 14–15 YEARS

Landmark 24

Calculate your ecological or carbon footprint and make a plan to reduce it.

Why?

For youth, learning about responsibility is very much about having the ability to respond. Youth are more likely to act when big global environmental problems are broken down into local and personal action. By calculating their ecological or carbon footprint, young people start to see how their everyday actions impact the planet and their own communities. When they collaborate with their friends and classmates, they can share ideas and support each other in making better choices for a healthier environment.

Students found their "Earth Overshoot Day" and calculated how many Earths would be needed to sustain their lifestyles. They then made recommendations on how they could personally reduce their carbon footprint.

—Karen Brown (teacher)

Climate Change Connections

Tackling climate change requires a combination of individual lifestyle changes (bottom-up) combined with progressive legislation and research (top-down). A good place for young people to begin is to personalize the issue by exploring how their lifestyles impact environmental health. By calculating their personal ecological or carbon footprints, they can think about solutions within their control. Some of these solutions might involve individual choices (such as which clothes to buy and how to travel), and others might require contacting decision-makers to advocate for policy changes. In either case, making a commitment to action, especially if undertaken with their peers, can be empowering and inspiring as young people transition toward adulthood.

How?

● **Calculate Your Own Carbon Footprint** by using an online calculator such as www.footprintcalculator.org. Another option is the International Student Carbon Footprint Challenge, which offers separate versions for middle school and secondary school at www.depts.washington.edu/i2sea/iscfc/calculate.php.

An **ecological footprint** is a measure of how much people take from nature compared to the amount of natural resources available for people.

Directions: Answer each section based on a typical day for you in your home and community. Record your scores on the next page.

Water Use

. My shower/bath on a typical day is:
   No shower/bath **[0]**
   3-6 minutes/half-full tub **[40]**
   10+ minutes/full tub **[70]**
. I flush the toilet:
   Every time I use it **[40]**
   Sometimes **[20]**
. I leave the water running when I brush my teeth. **[40]**
. We use a dishwasher at home. **[50]**
. We water our lawn at home. **[50]**

Food

1. I eat animal-based products:
   Never (vegan) **[0]**
   Infrequently (vegetarian) **[10]**
   Occasionally **[30]**
   Often (balanced) **[50]**
   Very often (meat daily) **[70]**
2. I compost my fruit and veggie scraps.
   Yes **[-10]**
   No **[20]**
3. _____ of my food is locally grown.
   All **[0]**
   Some **[20]**
   None **[40]**
4. _____ of my food is packaged.
   All **[40]**
   Some **[20]**
   None **[0]**
5. _____ of my food is processed.
   All **[40]**
   Some **[20]**
   None **[0]**

At Home

1. I live in a:
   Single-detached home **[200]**
   Townhouse **[100]**
   Apartment **[50]**
2. I have a second home, like a cotta that is often empty. **[200]**
3. We donate items when we no longe need them. **[-10]**

Energy Use

1. _____ of our appliances are energ efficient.
   All **[0]**
   Some **[20]**
   None **[40]**
2. I dry my clothes in a machine dryer. **[50]**
3. My house has compact fluoresce lights. **[-10]**
4. My house has air conditioning. **[5**

Transportation

1. I walk or bike to school. **[-20]**
2. The time I spend in a vehicle on typical day is:
   None **[0]**
   Less than half an hour **[20]**
   Half an hour to an hour **[3**
   An hour or more **[40]**
3. We have more than two cars. **[100**
4. I travel by plane at least once a year. **[200]**

Credit: Amanda Hipgrave

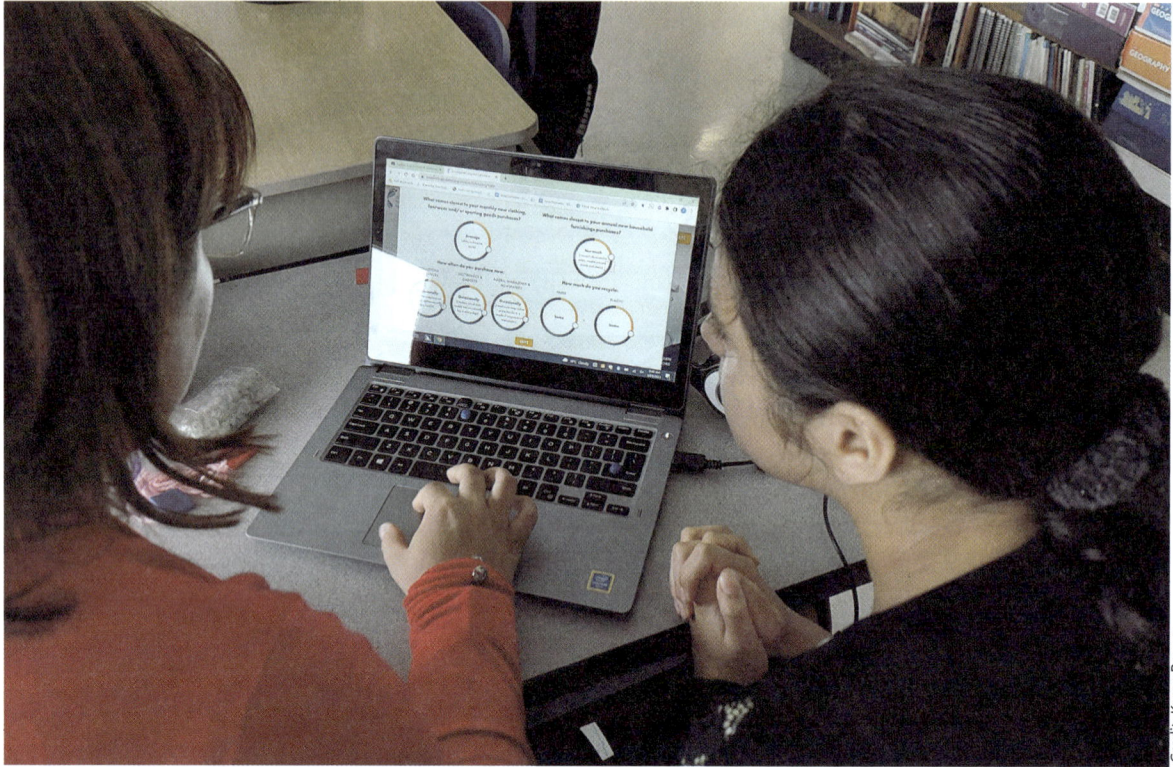

Credit: Karen Brown

Researching Earth Overshoot

Try to leave the Earth a
better place than when
you arrived.

— SIDNEY SHELDON (writer)

Unless someone like
you cares an awful lot,
nothing is going to get
better. It is not!

— DR. SEUSS, from *The Lorax*

● **Your Impact:** How did your results compare with the average footprint for your country and the global average? What might account for any differences? Consider which aspects of your lifestyle you can personally control, such as food choices, transportation methods, and energy use at home. Recognize that some factors, like the location of your home, may be beyond your control.

● **Reduce Your Carbon Footprint:** What are some reasonable changes you could make to reduce your footprint? Compare your ideas with those of other students and family members to find practical ways to reduce your footprint.

- **Lifestyle Choices:** What are ways in which you might influence the lifestyle choices of others in your community? Consider organizing awareness campaigns, creating educational materials, or starting community projects like carpooling or local cleanups.
- **Journal:** Keep a journal or create a digital log to track your progress in reducing your carbon footprint over time. Note the changes you make and their impact.
- **Presentation:** Develop a simple presentation or workshop to educate your peers and community members about carbon footprints and sustainable practices.
- **Eco-Friendly Challenge:** Organize eco-friendly challenges within your school or community, such as "no plastic week" or "bike-to-school month," to promote sustainable habits.
- **Carbon Offset Projects:** Research and participate in carbon offset projects, such as tree planting or renewable energy initiatives, to compensate for your footprint.
- **Sustainable Products:** Investigate and promote the use of sustainable products,

> **We are the first generation to feel the impact of climate change and the last generation that can do something about it.**
> —BARACK OBAMA (former U.S. president)

like reusable bags, eco-friendly cleaning supplies, and energy-efficient appliances.
- **Advocacy:** Consider participating in advocacy by writing letters to local representatives or participating in community meetings to support environmental initiatives.
- **Eco-Club:** Join or establish an eco-club at your school to collaborate on sustainability projects and share ideas on reducing carbon footprints.
- **Your Pledge:** Create a personal pledge or commitment to reduce your carbon footprint and encourage others to do the same. Display these pledges in a public space to inspire collective action.

> **Students calculated their ecological footprint, examined how Canadian lifestyles compare with other countries, and made personal plans to reduce our footprints.**
> —AMANDA HIPGRAVE (teacher)

AGES 14–15 YEARS

Landmark 25

Explore and develop at least three outdoor skills that are new to you.

Why?

Teens love testing their limits and figuring out who they are and where they fit in the world. At ages 14 to 15, many are drawn to the excitement of adventure and the thrill of trying new things. This is also a time when they're exploring independence and questioning the status quo. With all the pressures from friends, school, and family, this can be a stressful period. Exploring new outdoor skills can be a great way to relieve stress, build physical stamina, and take a break from screens. Trying out activities like hiking, canoeing, camping, or wilderness survival not only offers adventure but also helps teens develop resilience, confidence, and a deeper connection to nature.

Climate Change Connections

This Landmark provides another opportunity to expand life skills that don't require fossil fuels. Whether engaging in adrenalin-pumping excitement such as white-water canoeing or wind-surfing, or quieter challenges such as archery or astronomy, young people can build skill and self-esteem at the same time. Life extends far beyond the internet!

How?

● **New Adventure Sports:** Explore opportunities to try new adventurous sports such as rock-climbing, white-water canoeing, cross-country skiing, or kayaking.

● **Wilderness Trip:** With the help of an adult and peers, try an overnight canoe or wilderness backpacking trip.

● **Navigating:** Learn how to find your way in a natural area using maps, compass, and/or GPS. Get involved in orienteering, which combines running with navigation, using a map and compass.

● **Fire Making:** Try learning how to make a fire without matches or paper. Learn about fire safety and how to be responsible with fire when outdoors.

● **Night Sky:** Learn how to recognize at least two constellations in the night sky in each of the four seasons. Learn how to tell the four directions using clues in the sky. Learn some of the stories about constellations told by different cultures.

● **Survival Skills:** Take a course or read a book on basic survival skills such as building a shelter, finding water, and signaling for help.

Credit: Cam Douglas

**Being brave isn't the absence of fear. Being brave
is having that fear but finding a way through it.**
—Bear Grylls (survivalist)

I see my path, but I don't know where it leads.

Not knowing where I'm going is what inspires me.

—Rosalia de Castro (poet)

Credit: Cam Douglas

The most dangerous thing you can do in life is play it safe.
—Casey Neistat (filmmaker)

- **Geocaching:** Participate in a geocaching adventure to learn about using GPS technology to find hidden treasures.
- **Practice Wilderness First Aid:** Learn how to handle common outdoor injuries and emergencies through a wilderness first aid course.
- **Explore Birdwatching:** Get involved in birdwatching; learn to identify local bird species and understand their behaviors and habitats.
- **Local Conservation Group:** Join a local conservation group and help with projects like trail maintenance, habitat restoration, or wildlife monitoring.
- **Learn to Fish:** Take up fishing, understanding the techniques, equipment, and sustainable practices for catching and releasing fish.
- **Master Paddleboarding or Windsurfing:** Try paddleboarding or windsurfing on a local lake or river.
- **Join a Nature Photography Workshop:** Learn the basics of nature photography and capture the beauty of the natural world through a camera lens.
- **Engage in Outdoor Yoga or Meditation:** Practice yoga or meditation in natural settings to enhance mental well-being and connect with nature.

Credit: iStock

- **Predict the Weather:** Learn to predict the weather by understanding clouds, wind patterns, and other natural signs.
- **Edible Wild:** Only under the guidance of an expert, learn to safely forage for wild edibles in your area. Only ever harvest those plants you know for certain are safe to consume.
- **Tracking:** Study animal tracks and signs to learn how to find and observe wildlife in their natural habitats.

AGES 14–15 YEARS

Landmark 26

Volunteer to help in your community in at least three different ways. Reflect on what you learned through music, poetry, a blog, journal, or social media.

Why?

At this age, teens are discovering their passions and talents while developing a sense of responsibility. Volunteering is a powerful way for them to connect with their community, engage in meaningful impact, and acquire valuable life skills. By learning how to help by participating in local cleanup events, assisting at food banks, or mentoring younger students, teens can explore different interests and find what truly resonates with them.

We make a living by what we get, but we make a life by what we give.
—Winston Churchill
(former UK Prime Minister)

Reflecting on their experiences through creative outlets like music, poetry, blogging, journaling, or social media allows teens to process their thoughts and emotions, share their stories, and inspire others. These reflections can help them understand the importance of community service, recognize their contributions, and appreciate the diverse ways they can make a difference. Volunteering not only benefits the community but also empowers teens to grow as individuals, builds empathy, and fosters a lifelong commitment to helping others.

Climate Change Connections

We don't learn to read overnight, and neither do we learn to tackle big issues like climate change quickly. At an age when teens are generally focused on themselves and their peer group, they can broaden their perspectives and develop valuable skills by trying a variety of challenges that benefit the community. Whether directly involved in environmental work or volunteering in some other way, young people develop critical leadership and problem-solving skills through volunteering in different ways. Spending time with people of different ages, opinions, or backgrounds can play a role in learning patience and flexibility and making unlikely friends. These experiences also contribute to the knowledge that each of us can have a positive impact on our community.

How?

Discuss Local Issues and Find Solutions:

● Host discussions with your peers or organize a town hall meeting where community members can voice their concerns and brainstorm solutions.

● Join local youth advisory boards or councils to represent young people's perspectives on community issues.

● Use social media, posters, and local events to raise awareness about important local issues and encourage community involvement.

> **The best way to find yourself is to lose yourself in the service of others.**
>
> —MAHATMA GANDHI (Indian lawyer)

Work with a Local Conservation Organization:

● Participate in or organize planting events to help restore forests, grasslands, and wetlands and create more diverse green spaces. Help with the aftercare of these projects.

● Join or lead clean-up efforts in parks, rivers, and other natural areas to remove trash and invasive species.

● Collect data on local wildlife, plants, or water quality to help conservation organizations monitor and protect ecosystems.

Credit: Karen Brown

Exploring what it's like to be blind

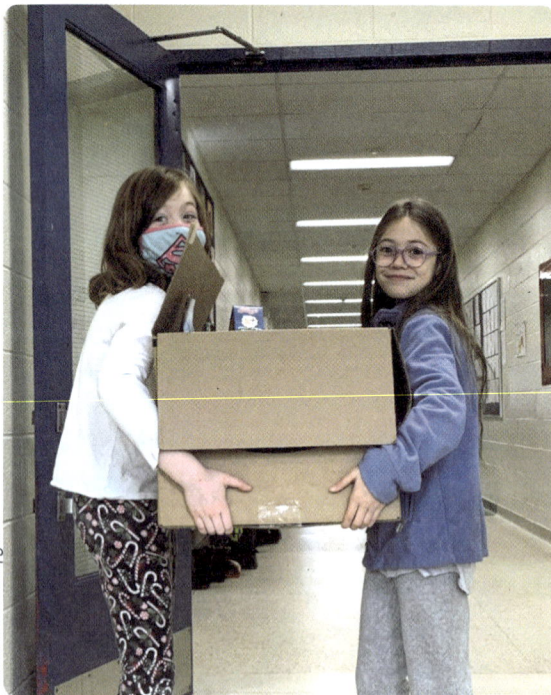

Credit: Amanda Hipgrave

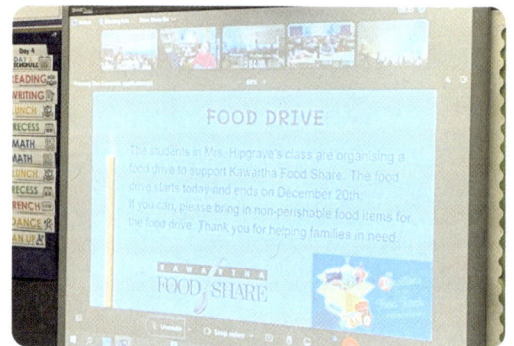

Over the years we've learned a lot from, and about, our classmate Israa. Israa is blind and has to learn in ways that are different from the rest of us. In addition, she does it in her second language as Arabic is her first. With several Outdoor Ed trips ahead, we thought it might be time to learn a bit more how she navigates. We needed a better understanding since we will be hiking in terrain that is unfamiliar to her. We hope we will be better guides and helpers in the future.

—Karen Brown (teacher)

Help at a Local Food Bank or Food Co-operative:
- Volunteer to prepare and serve meals at soup kitchens or community meal programs.
- Help to create a small community garden to supply food banks with healthy, local produce.

Help at a Local Animal Shelter:
- Volunteer to feed, groom, and play with animals to keep them healthy and socialized.
- Assist with organizing and running events to help animals find forever homes.
- Help to find temporary homes for animals that need special care or are waiting for adoption.

Conduct a Food Drive:
- Engage schools, local businesses, and community groups to collect non-perishable food items.
- Help to organize themed food drives around holidays when the need for food donations often increases.
- Educate your peers and community about food insecurity and the importance of supporting local food banks.

Conduct a Fundraising Project:
- Help to organize sponsored runs or walks to raise funds for local environmental or social causes.
- Use crowdfunding platforms to gather donations for organizations you admire.
- Host bake sales, car washes, or talent shows with proceeds going to support a chosen charity.

Help with Transportation Surveys:
- Volunteer to gather data on local transportation habits, such as counting bicycles, pedestrians, and public transport users.
- Use the data collected to advocate for better infrastructure for active transportation, such as bike lanes and pedestrian paths.
- Organize workshops to educate the community about the benefits of active transportation and how they can contribute to healthier, more sustainable cities.

Our students are *amazing* volunteers. They help our primary classes every day by acting as lunch monitors in various classes and playing with primary students at recess. It is wonderful to watch them make such positive contributions to our school community.

—AMANDA HIPGRAVE (teacher)

Volunteer at a Local Seniors' Residence:
- Provide companionship to seniors who may be experiencing declines in physical or mental capacities.
- Share your skills in using technology such as computers and internet.
- Did these seniors grow up in this area? Where are they from and how has the world changed during their lifetimes?
- Can these seniors teach you games they enjoy playing?
- Talk about the things that are important to each of you, and how you enjoy spending your time.

Volunteer at a Local Daycare or Early Learning Center:
- Young children love spending time with teens, who can often become "heroes" to them!
- Share books together, explore outdoors, talk together, and be a good listener.
- What new skills are children trying to master at this age? Are you able to help with this?
- Be sure to ask the program supervisor for suggestions on how you can be a good mentor to these young children.

© sunet / Adobe Stock

Landmarks for Age 16 to 17 Years

**Age 16 to 17 Years:
Characteristics of This Age Group**

During the mid to late teen years, young people are driven by a powerful desire for independence and self-discovery. This is a time of emotional intensity, as teens grapple with deeper questions about their identity and future. They have a heightened sense of justice and are highly attuned to social issues, which can fuel their passion for activism and change. Peer influence continues to play a significant role in their lives, and they may begin forming longer-term romantic relationships.

Credit: Cam Douglas

Teens at this age are energized by both physical and mental challenges, and they can exhibit remarkable dedication to causes they believe in and support. Their developing sense of compassion extends beyond themselves to others, as they become aware of the interconnectedness of their actions and the broader world. This compassion often translates into a strong commitment to the well-being of others, including an increasing concern for environmental sustainability, social justice, and political issues.

Older teens are often preoccupied with thoughts about the future and their role within it. Providing them with opportunities to make meaningful contributions to their communities can be empowering and help them channel their concerns into positive action. They are also natural leaders, capable of guiding and inspiring younger children in areas such as sports, arts, academics, and recreation. As they build self-confidence and expand their understanding of the world, they begin to recognize the unique talents they possess and the ways they can make a real and lasting difference in their community and beyond.

AGES 16–17 YEARS

Landmark 27

Plan and go on an extended trip in a wilderness area for at least 3 to 5 days. Travel by canoe, bicycle, skis, on foot, snowshoe, or any other self-propelled mode of travel.

Why?

Persevering, even when things are tough, shows us that we are stronger than we think we are. Feeling the support of our friends, facing challenges and overcoming them are lessons that will serve us throughout our lives. Advanced outdoor experiences (including planning, leading, and evaluating) help our youth to acquire leadership skills, deal with conflict, and learn about teamwork and the roles that people play within a group. Surviving and thriving in unfamiliar and sometimes uncomfortable circumstances helps to build self-esteem and strengthen a sense of identity and accomplishment. Encountering the challenges of living outdoors for an extended time can be a life-changing experience.

Climate Change Connections

For so many people, an extended wilderness trip provides an unparalleled opportunity to learn what is really important in life. You experience the power and beauty of the natural world first-hand and realize the importance of teamwork in meeting new challenges. The worries and stresses of life at home fall away as you experience, perhaps for the first time, the reality of living on the land. This provides a stark contrast to a consumer-based world and stimulates a great deal of reflection about choices and priorities.

How?

● **Planning:** Help to plan a family wilderness adventure and take on a leadership role in organizing the trip. Make sure someone else knows when you're leaving, where you're going, and when you plan to return. If you're travelling in a public park, the staff will take this information.
● **Wilderness Program:** If you can, enroll in a summer camp or a program that specializes in wilderness tripping, providing structured yet adventurous experiences. If your school has one, enroll in secondary outdoor education programs that include extended wilderness trips as part of the curriculum.

● **Menu Planning:** When planning menus, research and plan nutritious, lightweight, and easy-to-prepare meals. Include dehydrated and nonperishable food options.

● **Essential Gear:** Learn about essential gear. Think about the importance of packing light and carrying only necessary items. Include a first aid kit.

● **Being Present:** Try to put your electronic device away in a pack and fully immerse yourself in nature. Use this as an opportunity to practice mindfulness and observe the world around you more closely.

In every walk with nature, we receive far more than we seek (paraphrased).
—JOHN MUIR (conservationist)

It takes courage to grow up and become who you really are.
—E.E. CUMMINGS (poet)

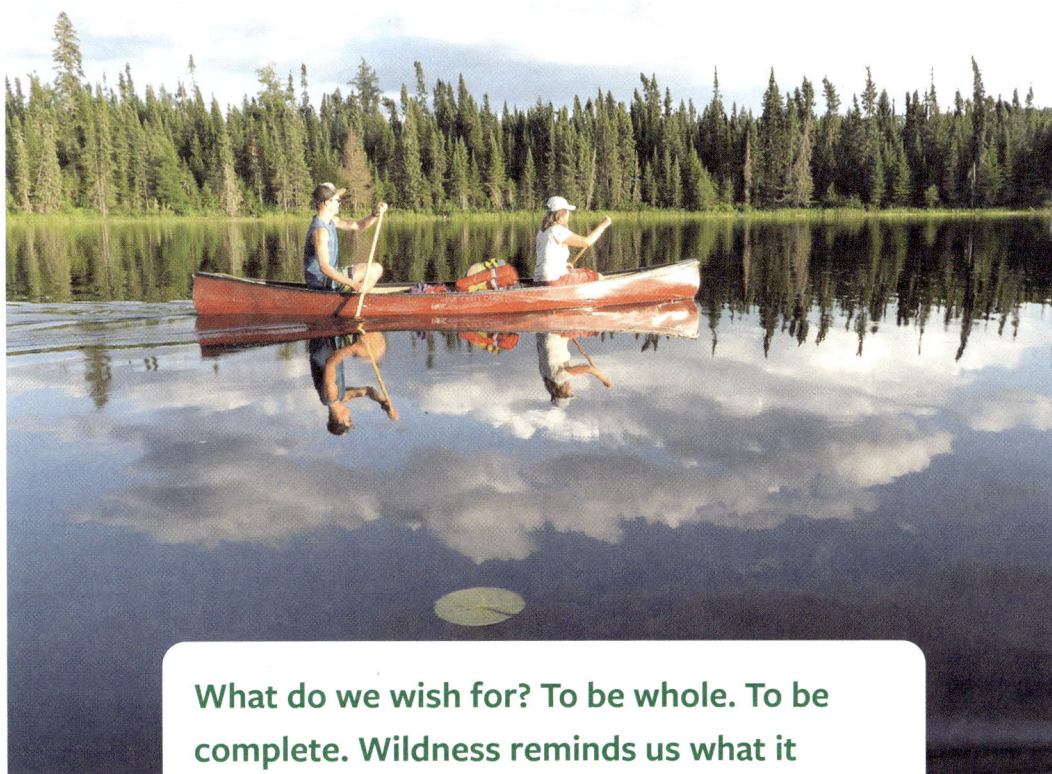

What do we wish for? To be whole. To be complete. Wildness reminds us what it means to be human, what we are connected to rather than what we are separate from.
—TERRY TEMPEST WILLIAMS, in *Red: Passion and Patience in the Desert*

Credit: Cam Douglas

• **Journaling:** Bring along a journal. Write about your experiences daily or after you return. Reflect on how this experience makes you feel, new things you learned, new skills you acquired, and how this trip has made a difference in your life. Consider also documenting your journey through sketches, musings, thought bubbles, and/or poetry.

• **No Trace:** Practice "Leave No Trace" principles to minimize your environmental impact.

• **Survival:** If you can, learn about basic wilderness survival skills such as building shelters, foraging for food, and purifying water. Learn simple navigation skills using maps and compasses and understand how to read natural signs for direction.

Credit: Cam Douglas

Our family spent a glorious eight days of canoeing through Quetico Provincial Park. The park was virtually empty, and we went almost four days without seeing a soul. Breath-taking scenery, gorgeous campsites, spectacular sunsets and the loons completely outdid themselves with their evening symphony. Such a great way to recharge the batteries. Just us, together, with our two canoes, tents, food, and great sense of adventure. The days are hard work, with the paddling, the wind, and especially the portaging. But all that builds one's sense of confidence— internally, and in a sense of wilderness skills.

—CAM DOUGLAS (parent and teacher)

- **Practice mindfulness.** Watch the stars at night. Listen to the natural sounds around you. Pay attention to the various shades of green or brown. What natural smells surround you? Try to empty your mind and really be present in this place.

- **Indigenous Connections:** Research and explore the historical and cultural significance of the wilderness area you are travelling through, learning about the Indigenous People whose cultures developed here. What can you learn from them in how they relate to land and their ethic of stewardship?

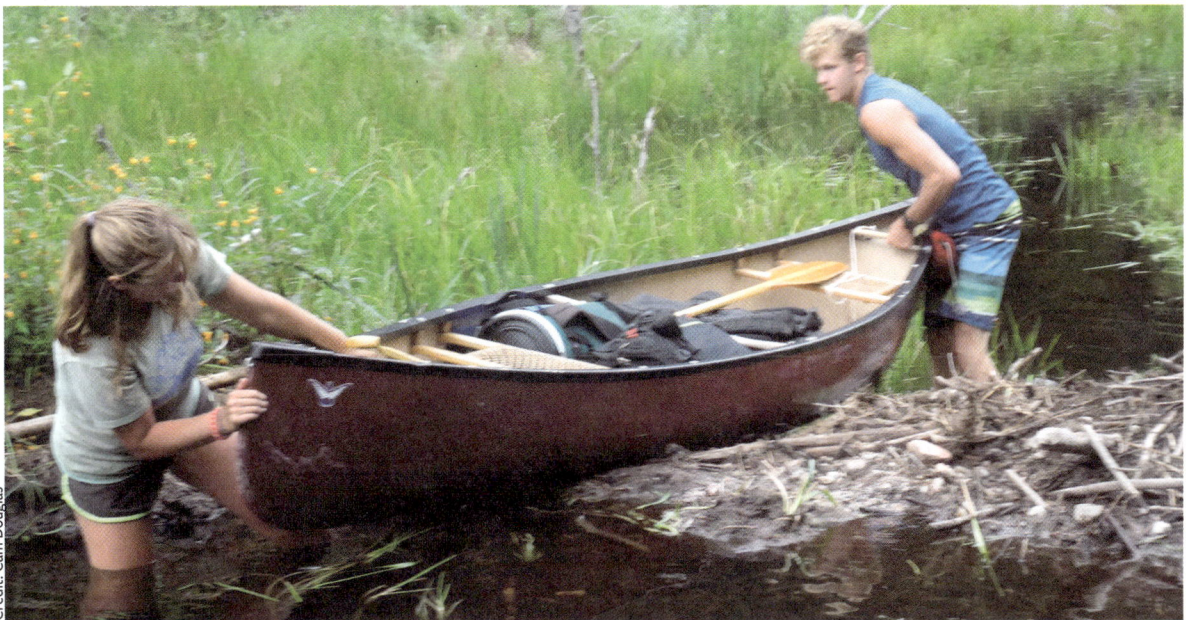

Credit: Cam Douglas

AGES 16–17 YEARS

LANDMARK 28

Help to rehabilitate something that has been damaged (such as an animal, waterway, or natural area) over an extended period of time.

Why?

From climate change to pollution and habitat destruction, these are daunting problems we are leaving for our children to solve. It is so important that we activate, motivate, and inspire our youth to do something positive for the Earth. And that can start right here—in the very place we occupy. Let's help our youth find simple ways to rehabilitate, restore, and bring back health and vitality to our communities. That can involve working with people, animals, or natural areas, depending on your personal skills and interest. A sense of hope and empowerment can grow from knowing that people are a vital healing and restoring force in the world. Through these hands-on experiences, young people can witness the direct results of their efforts, helping to spark a lifelong commitment to environmental stewardship, ensuring a healthier, more sustainable future for all.

Climate Change Connections

An important concept in climate change mitigation is carbon sinks. Plants naturally pull carbon out of the air as part of photosynthesis and store it in their tissues. This storage is called a sink. While considering options for environmental rehabilitation, think about increasing the volume of plant tissues (biomass) in a degraded area. Trees are especially good for this since they add to their volume of wood each year. Include fast-growing trees such as poplars, as well as slow-growing, long-lived trees such as oaks as climate-conscious choices for habitat rehabilitation. In grasslands, plants such as *Panicum* sp. accumulate a great deal of biomass each year. Check with local experts to see which species are appropriate for your project. Remember that planting is only one step in the process of habitat restoration. Follow-up maintenance and monitoring are equally important.

How?

- **Wetland Restoration:** Partner with local environmental organizations to restore a damaged wetland. Participate in planting native vegetation, removing invasive species, and monitoring water quality over time.
- **Stream Bank Stabilization:** Work with a conservation group to stabilize eroding stream banks. Plant deep-rooted native plants, install bioengineering structures, and help reduce sediment runoff.
- **Urban Tree Planting:** Join a tree-planting initiative in your community to replace trees lost to disease, development, or extreme weather. Help care for the trees during their critical first years.
- **Invasive Species Removal:** Identify areas in your community where invasive plant

When enough people come together, then change will come, and we can achieve almost anything. So instead of looking for hope, start creating it.
—GRETA THUNBERG (climate activist)

The students from a grade 11 Environmental Science class from Holy Cross planted trees in Ennismore with Otonabee Conservation as part of the rehabilitation of a gravel pit.
—MIKE HALLORAN (teacher)

Credit: Cam Douglas

Credit: Mike Halloran

We enhanced the riparian buffer along a watercourse.
—Amanda McInnes (teacher)

species are threatening native ecosystems. Organize regular volunteer events to remove these species and allow native plants to return.

● **Pollinator Habitat Restoration:** Transform neglected or abandoned lots into pollinator-friendly habitats by planting native flowers and providing nesting sites for bees, butterflies, and other pollinators.

● **Stream or River Monitoring:** Adopt a local stream or river to monitor over time. Collect data on water quality, observe wildlife, and take action to address pollution or erosion.

● **Rewilding:** Talk to school administrators about rewilding a section of your schoolyard. Create a small pollinator garden; plant wildlife-friendly trees and shrubs. Learn how to care for them.

● **Measuring Carbon in a Tree:** Trees are amazing carbon sinks. Using a tree carbon calculator available online, find out how much carbon is stored in the trees around your school grounds. How much more could be stored by planting native species after 5 years, 10 years, 100 years? Imagine if every school in your community dedicates part of their yard for more trees, how much carbon could they collectively store?

● **Local Wildlife Rehabilitation:** Local wildlife rehabilitation centers may have volunteer opportunities to care for injured wildlife while they recover and prepare for release back to their homes.

● **Animal Shelters:** Local animal shelters may have opportunities to care for and socialize with pets who are waiting for adoption.

● **Community Cleanup:** Plan and lead community cleanup events in local parks, rivers, and beaches. Remove trash, plant trees, and restore natural habitats.

- **Advocacy:** Collaborate with local government and community groups to advocate for the creation and preservation of green spaces and natural areas in your community.

- **Support a Friend:** Do you have a friend or relative who's recovering from an accident or illness? Can you provide support for them during their recovery? You could help in a variety of ways such as providing companionship, preparing and delivering a meal, making a cup of tea, playing games together, tidying up their house or yard, or reading stories from a book or the internet.

> ## Worldwide, we have now destroyed over half of the forest that once flourished on our planet. Not only are we losing the animals that once lived on them, we're also changing the climate of the entire globe.
> —DAVID ATTENBOROUGH (broadcaster and biologist)

Credit: Mike Halloran

AGES 16–17 YEARS

Landmark 29

Explore a local issue of social justice. Develop a plan to raise public awareness or motivate public involvement.

Why?

This Landmark continues a focus on social issues at the stage of life when social interactions are such a dominant concern. Teens are encouraged to identify and discuss a local issue of social justice that is meaningful to them. Young people benefit from opportunities to explore issues such as racial or religious intolerance, gender identity and discrimination, poverty, and environmental racism. Discussion of issues should be linked to exploring potential solutions, and opportunities to take meaningful local action. This is an important opportunity to develop and practice leadership skills and expand an understanding of the power of collective action.

Climate Change Connections

A fascinating study from the UK, by Howell and Allen, reveals that the drive for social justice can be an even stronger motivator for climate action than environmental concern alone. For many, caring about human well-being is a natural first step, which can later evolve into a deeper

commitment to protecting our planet. This is especially vital for those who didn't grow up deeply connected to nature. The harsh reality is that the world's most vulnerable populations are often the ones who bear the brunt of climate change. The health of our planet and the well-being of humanity are inseparably intertwined; when we fight for one, we fight for both.

How?

● **Local Issues:** Identify issues of injustice that concern the students. Have they experienced or witnessed examples of social discrimination? What may be the root causes of these issues? What role can they play in promoting justice and equity?
● **Homelessness:** Is homelessness an issue of concern in your community? Have students met or talked with someone experiencing homelessness? What factors may contribute to this situation? How could students raise public awareness of local homelessness and encourage public participation is addressing the issue?

Our history is a living history, that has throbbed, withstood and survived many centuries of sacrifice. Now it comes forward again with strength. The seeds, dormant for such a long time, break out today with some uncertainty, although they germinate in a world that is at present characterized by confusion and uncertainty.

—Rigoberta Menchú Tum (K'iche' Guatemalan feminist, activist, and Nobel Peace Prize Laureate)

We hosted an interactive workshop on Youth as Agents of Change in the Sustainable Development Goals (SDGs) to engage students in social justice issues, including poverty, climate change, racial discrimination, and leaving no one behind. During the workshop, we encouraged students to participate in discussions by asking a series of thought-provoking questions. We explored examples and useful case studies highlighting the underlying factors causing poverty in Peterborough and local actions taken to address these issues. We spoke about how the local community can work toward resolving these issues and how youth can take action and make positive change to build a more sustainable future.

—Margaret Zou (community educator)

● **Mind Mapping:** Make a mind map of the issue that concerns you. Explore the various root causes of the issue, including ideas from other people. Add to your map any actions you can think of to help address the issue.

● **Share Your Concerns:** Express and share your concerns and ideas for action to people in positions of influence such as politicians, school board officials, directors of social or environmental organizations.

● **Positive Action:** Identify, celebrate, and promote examples of positive action being taken to find solutions to issues that concern you. Use social media platforms to spread awareness about the issue. Create posts, videos, or infographics that highlight key facts, personal stories, and ways for others to get involved. Develop pamphlets, posters, or presentations to educate others about the issue. Share these materials with schools, libraries, or community centers, helping to spread knowledge and inspire action.

● **Hammond's Action Triangle:** Bill Hammond once said that young people need to be provided with strategies that help them engage in positive action. His advice? Learn about action, participate in action, and reflect on action. Select one local social justice action that is important to you. Develop some advocacy skills: how to write, how to speak, and how to promote your cause. Create a small

We discussed with students how the United Nations Declaration on the Rights of Indigenous Peoples (UNDRIP) principles would ensure that Canada meets the universal framework of minimum standards for the survival, dignity, and well-being of the Indigenous Peoples, and a stronger degree of protection for Indigenous rights than those currently in place under Canadian law. We engaged students throughout the workshop using interactive activities on Slido and Kahoot. Students were encouraged to participate in reflections and discussions on how UNDRIP resonated with them and how their knowledge and understanding of UNDRIP can help to generate a positive impact in the local community.

—Margaret Zou (community educator)

campaign. Reflect on what went well and what you would change for next time. Effective action is a skill which can be refined over time.

● **Raising Awareness:** Organize a forum or panel discussion with local experts, activists, or community members affected by the issue. Select a film or book that addresses the social justice issue and organize a screening or discussion group. This can be a powerful way to spark conversation and deepen understanding of the topic.

● **Collaboration:** Collaborate with local nonprofits, advocacy groups, or community organizations already working on the issue.

● **Local Policies:** Research local policies related to the issue and advocate for change. This might involve writing letters to elected officials, attending town hall meetings, or even starting a petition to gather community support.

● **Art and Expression:** Create murals, sculptures, or other visual art pieces that reflect the issue and convey a message of hope or change. These artworks can be displayed in public spaces to raise awareness and encourage public involvement.

AGES 16–17 YEARS
Landmark 30

Use a creative medium to describe your "ecological self." How are you are connected to the world? Express your perspective in poetry, music, visual or dramatic art.

Why?

A fundamental concept in Indigenous teachings is that we are all interconnected. This knowledge of the inextricable links between all beings also brings a great deal of responsibility. Other beings support us, and we must in turn support the well-being of life on Earth. As young people shape their lives and grow into adulthood, they can find comfort and purpose in the knowledge that they are part of a large, complex, and interwoven family of life on Earth. This awareness grounds and centers us and can be a lifeline amidst the stress and complexity of modern life. We feel more complete and less alone when we recognize that we belong to something much bigger than ourselves.

Climate Change Connections

The phenomenon of global climate change is a perfect example of how our life choices have changed the composition of the atmosphere, which is now resulting in extreme weather events including droughts, floods, windstorms, and wildfires. At the same time, we all benefit every day from many gifts from the Earth: food, shelter, beauty, companionship, and more. Taking time to think about the many ways we impact, and are impacted by, the world we live in helps to expand our sense of time and shape our identity. Haudenosaunee teachings remind us that our lives are the result of decisions made by seven generations before us, and that we must make decisions today for the well-being of seven generations to come.

How?

● **Your Ecological Self:** What is the meaning of "your ecological self"? This involves thinking about all the ways that you are connected to the world around you. Your human community is certainly an important part of that, but you also depend on air to breathe, water to drink, and many other

Each of us is a unique strand in the intricate web of life and are here to make a contribution.

—Deepak Chopra (author)

living things that provide your food and shelter. In turn, you also affect your world in many ways. Your personal ecological self is like a fingerprint: it is unique to you, and unlike anyone else's.

● **Representation:** We encourage you to represent your ecological self in any way you wish: through prose, poetry, a drawing, a sculpture, a journal entry, a video, or anything else you choose. In many ways, this is the pinnacle of the journey toward stewardship and kinship—an awareness of the incredible tapestry of life where each of us is an important thread and plays a special role in holding the fabric together.

> **Those who contemplate the beauty of the Earth find reserves of strength that will endure as long as life lasts.**
>
> —RACHEL CARSON (author)

Credit: Emily Warren

- **Storytelling:** Create a short story or narrative that explores your connection to the natural world. Imagine how the landscape, animals, and plants around you shape who you are and where you belong.
- **Nature Collage:** Gather natural materials like leaves, twigs, stones, or shells and use them to create a collage that symbolizes your relationship with the Earth. Consider what each element represents in your ecological self.
- **Environmental Mandala:** Design a mandala using natural elements that represents the harmony and balance in your ecological self. It may be helpful to research how mandalas have been used by different religions to represent their core beliefs. Think about how the different components of life are interconnected and essential to your well-being.
- **Musical Composition:** If you are musical, compose a piece of music that reflects your relationship to the world around you. Use sounds from the environment as inspiration or even as part of the composition.
- **Eco-Theater:** Write and perform a short play or skit that dramatizes your ecological self. Explore how your actions and choices impact the environment and how the environment, in turn, shapes you.
- **Photography Project:** Create a series of photographs that capture your interactions with your world. Highlight moments where you feel most connected and alive.
- **Reflective Writing:** Keep a reflective journal that explores your thoughts, feelings, and experiences of the many things that affect your life. Describe how these experiences shape your understanding of your place in the world

May your trails be crooked, winding, lonesome, dangerous, leading to the most amazing view. May your mountains rise into and above the clouds. May your rivers flow without end.

—Edward Abbey (author)

It Takes a Village

We're energized and inspired by all the wonderful people we see every day, doing great work with kids! We want to see every young person being supported and encouraged in their journey through life—to truly feel the gift of being alive and look to the future with excitement and resolve. We live at an unprecedented time in human history, where we literally have the power to destroy the very systems that support life on Earth. Or, with a tremendous collective effort, we can create a different story—of hope, healing, and valuing life and health above all else. The Pathway to Stewardship and Kinship is one way to practice that shift in values, by guiding every young person to cultivate life-sustaining relationships with the land and their community. And, in the process, we as adults, can learn and grow along with them.

While the Pathway program was developed and tested in the community of Peterborough, Ontario, Canada, it could easily be adapted to be applied anywhere. Every child is unique, but children follow the same basic patterns of growth and development everywhere in the world. The 30 Landmarks can provide a simple starting point for nurturing stewardship wherever you live, with plenty of flexibility for adapting to many different cultures and landscapes.

The Pathway concepts can be applied individually by anyone spending time with children (parents, educators, caregivers, etc.), but its impact will increase dramatically when adopted by an entire community. Our goal in Peterborough was to work together, as a community, to infuse these concepts into our culture, so they would become a familiar part of everyday life. That seemingly lofty goal becomes much more attainable when broken down into a series of manageable steps. In this concluding chapter, we'll describe our process in more detail, with advice on the underlying attitudes and principles that are important ingredients for success.

Bring Together Diverse Perspectives with a Common Goal

A community is made up of many different people with differing backgrounds and perspectives. While we tend to gravitate toward people who think the way we do, true community-building involves reaching out and learning to work with people from many sectors—all of whom have unique skills and wisdom to share.

A powerful starting point is to focus on a goal you all have in common. In our case, our shared focus was children's health and well-being. We wanted to provide experiences that would help build personal and collective health in young people from birth through their teen years. That goal easily appealed to many people, and we made a conscious effort to reach out to as many different sectors as we could.

As well as conducting comprehensive research on best practices in environmental education and child development, we also interviewed 80 community leaders with an interest in the environment. We wanted to ask about the kinds of childhood experiences that stood out in their memories, helping to develop their sense of connection to the natural world and concern for protecting it. We included health professionals, educators, environmentalists, artists, faith communities, Indigenous leaders, psychologists, university students, parents, grandparents, youth leaders, and politicians. Later, as we convened a steering committee to guide the project, we made a point of including people with on-the-ground experience (teachers, caregivers, early childhood educators, parents), as well as influential community decision-makers who could help open doors as we progressed (school board consultants, principals, public health supervisors, leaders of environmental agencies, etc.).

The Landmarks of the Pathway program can also be a helpful guide for smaller groups, as well for families in a neighborhood, leaders of an after-school program, or any other collective of people working with children. Regardless of your financial situation, whether you live in the city or country, whether you're a long-time resident or recent immigrant, the Pathway program can be a rallying point for working together toward a common purpose. As the saying goes, "It takes a village to raise a child," and this program can benefit all children, in any corner of the world, if organizers stay focused on the common goal.

Establish a Sense of Shared Ownership

People feel more commitment to something they feel belongs to them. The Pathway Landmarks, supports, and outreach mechanisms were developed through extensive community consultation. They can be used as a starting point for your community, but there is plenty of potential to shape the program to suit your regional and cultural needs and strengths.

For example, Landmark 7 focuses on sharing nature-based songs, stories, poems, and games with children roughly 4 to 5 years old. This is a perfect opportunity to draw on materials from the cultures in your community. Local Indigenous cultures are rich storehouses of songs, stories, and games that promote living in relationship with the land, and these can be shared with children starting at a young age. Similarly, immigrants in your community bring their own treasure trove of materials that celebrate the place they came from. Honoring the natural world from a variety of perspectives reinforces a sense of expanded community, and time-tested regional wisdom provides a solid foundation to help children navigate through the confusing maze of modern life.

The suggested activities in the preceding Landmarks section of this book are only a starting point for stimulating your community's own ideas. By expanding the program with your own regional ideas, and using your unique resources, you help to create and maintain a sense of shared ownership, which is vital to ongoing success.

In large multi-sectoral programs, it is often practical and essential to have one or two organizations play the role of a backbone organizer—managing finances, assuming legal responsibility for the collective, and organizing meetings and working groups. However, be sure to reinforce the concept of shared ownership and responsibility wherever you can. If this becomes "somebody else's" program, you lose momentum and commitment from the group.

Build Respectful and Energizing Relationships

It's wise to start discussions with a group of people with whom you have a strong, supportive relationship, and work out from there into the broader community. This can help to create an atmosphere of cooperation and creative thinking. Building mutually respectful relationships takes time and patience but is well worth the effort!

Learning to be a good listener is absolutely essential in developing strong working relationships. The more you understand the needs and concerns of others, the

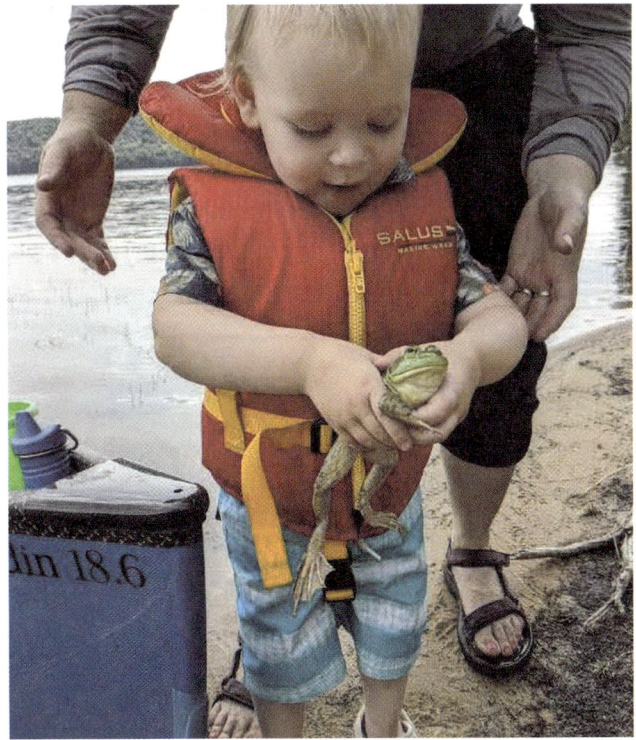

Credit: Sarah Parna

easier it will be to develop action plans that benefit everyone. This is especially important as you reach out to people with different views and priorities than your own.

In developing the Pathway to Stewardship and Kinship program, we created draft Landmarks for various age groups, based on the results of our 80 interviews with community leaders. Then, we convened a two-hour workshop to which we invited more than 100 representatives from a wide variety of community groups, all of whom had some involvement in working with children and youth. After a brief introduction to the Pathway program, we split the large group into smaller discussion groups, bringing together people of similar interests to discuss a few guiding questions. For example, we grouped together teachers and school board reps; artists and cultural reps; people from Indigenous communities; early childhood educators and parents of young children; reps from nonprofit groups; secondary school students. The smaller, more intimate discussion groups created a safe, friendly space for sharing a wide variety of perspectives. Each group had a discussion facilitator and a notetaker, as all the groups explored barriers and opportunities for working with the Pathway framework, gave feedback on the draft Landmarks, and discussed resources they needed or could offer. We then compiled and distributed summaries to all participants from each group's discussion, and used the feedback to further refine the program.

Seeing the program plans improve and strengthen as a result of collaboration was energizing for everyone! When participants could see that their voices had been heard and valued, they were more willing to step forward when it was time to put the plans into action.

Again, building positive and respectful relationships with others is equally important whether you're working on a community-wide basis or with a much smaller group to nurture stewardship and build health in the young people in your lives.

Assemble Local Talents and Resources

Every community has its own unique blend of talented and dedicated people. What skills and resources already exist in your community to help young people connect with and enjoy the natural world around them? Can you think of artists, educators, naturalists, outdoor sportspeople, writers, storytellers, musicians, biologists, gardeners, or others who are able to share their passion for the land in an engaging way for young people? It's ideal if you can access funding opportunities to allow you to pay these wonderful people to do Landmark-related programming with children and youth, or to provide training for educators and parents.

In our case, funding from the Ontario Trillium Foundation allowed us to pay quite a few local people to do outreach programs in schools and for families. These included:

- Several outdoor educators who engaged elementary students in Landmark-linked programs in schoolyards and nearby parks; some also conducted training workshops for teachers or parents.
- Water quality testing for secondary students was led by our local conservation authority; they also directed several habitat restoration projects with elementary and secondary students.
- A local musician and storyteller conducted nature-based interactive programs with preschool children and their parents.
- A local World Issues Center engaged secondary students in exploring the United Nations Sustainable Development Goals and developing local action plans.
- Our local zoo worked with elementary classes to virtually adopt a zoo animal which they visited weekly through Zoom, and learned the animal's needs and habits;

the zoo also offered a limited number of live animals for short-term loan to kindergarten classes, so the students could participate directly in caring for their foster pet (instructions, food, and materials provided by a zookeeper).

• Our local Indigenous education collective visited school groups with programs that combined Indigenous Knowledge with Euro-Western science.

• A local musician and puppeteer enthralled early learning groups with music, stories, and puppetry to share the wonders of the natural world.

• A local Indigenous educator delivered programs about Anishinaabe teachings and outdoor learning for families.

• Our local organization that welcomes and works with immigrant families visited elementary classes to share their stories of making a home in a new place.

• A local environmental organization organized bicycle workshops in local schools and for several First Nation communities.

• A nearby farm collective created a film on natural fibre production in local farms which they screened in classrooms. Students then chatted through Zoom with one of the farmers in the film.

• A local pollination expert worked with a whole school to learn about the importance of pollinators and to develop a schoolyard pollinator garden.

• A local turtle rescue center conducted virtual tours of their center for elementary classes, to teach about the importance of turtles and what children can do to protect them.

• Similarly, a local wildlife rescue center visited classes to teach about local wildlife and the role people can play in protecting them and their habitats.

• Local naturalists also volunteered to lead nature walks with elementary classes, exploring their local "neighborwood."

• Local Master Gardeners also volunteered to help school groups develop schoolyard food or habitat gardens.

Engaging local talent puts your own unique spin on Pathway Landmarks, and again reinforces a sense of shared ownership in the program.

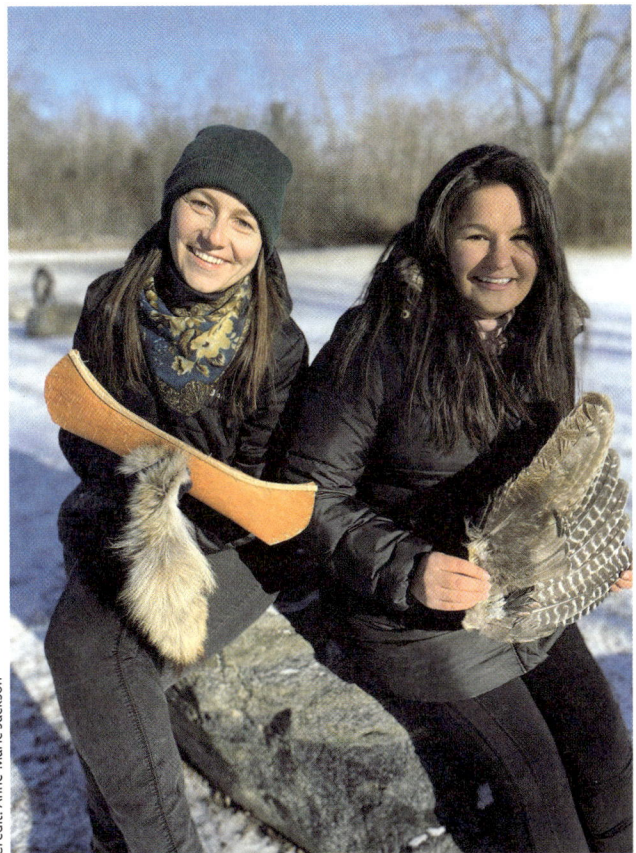

Credit: Anne-Marie Jackson

Track Progress

When we initially pilot-tested the Pathway program, we found it challenging to know if or when our participants were actually doing any Landmark activities. We had provided plenty of tools, but busy teachers and parents rarely responded to emails, or told us how they were progressing.

We decided to set up a simple reporting system to help us track participation, both for reporting purposes to our funder and to give us a window on the activities taking place in the community. This also allowed us to share ideas with others. Since many people are familiar with posting highlights of their activities on social media such as Facebook and Instagram, we set up a similar simple process for families and educators to quickly submit a few lines about their Landmark-linked activities, along with a photo. We linked this with the Pathway website, with a banner on the homepage for simple instructions on how to post a report.

Based on feedback from participants, we also created a simple website-based app that people could use with their cellphones, to quickly send in reports from anywhere. We asked for permission to share reports, along with photos, as part of the reporting process. If permission was given, we posted reports on our website, grouped roughly by age group. This gave participants a chance to see themselves online (like social media) and opened the door to seeing what others were doing with similar-aged kids. This all helped to reinforce a sense of community involvement, while keeping our finger on the pulse of participation.

Credit: Giselle Peters

> The video tutorial was so helpful, and once I started uploading our projects, it was so easy. I was overwhelmed previously, thinking it was so much more work than it actually is. I now realize we are hitting landmarks regularly with our programming.
>
> —(early childhood educator)

We offered a number of incentives to send in Landmark reports. We asked participants to commit to submitting at least one Landmark report each season, as a condition of receiving a variety of perks such as classroom visits from local guests. We also held a monthly draw, where anyone who had submitted a report that month could win a $50 gift certificate from a local business.

At the start of our three-year grant to roll out the program community-wide, we set a goal of documenting 10,000 Landmark experiences and crossed our fingers that we could generate that much activity. To our amazement, participants were so enthusiastic, and the incentives we offered were so successful, that we sailed past 10,000 in the first year after setting up the reporting process. After three years, we documented 62,591 Landmark experiences from local children and youth! That's a lot of good news stories to share with others, and impressive numbers for further expansion of the program.

Build Positive Momentum

Keeping up positive communication between participants and organizers was a critical factor in building momentum as we progressed. For anyone undertaking a similar program, we highly recommend implementing as many communication tools as your budget will allow. In our case, we were fortunate to have a generous funding base that enabled quite a few different kinds of communication between participants. We also had a great stroke of good luck in being able to hire a professional videographer to be our Communications Coordinator. This was especially helpful as we were rolling out the project during the COVID-19 pandemic, which limited opportunities for face-to-face communication.

We found all of the following tools for communication very useful in implementing the Pathway program. Your community may have other ideas (or a different budget), so using the skills at hand is always the best choice.

● **Websites:** Websites have become common communication tools, and many useful resources are now available to help even novices create and maintain simple websites. The Pathway website (http://pathwayproject.ca) has been an important hub for all Pathway activities as the project developed and grew. A website can be used to promote upcoming events, showcase activities being done in the community, and share a wide range of resources. On our website, we also posted the publications we had produced, including the results of extensive community interviews, curriculum links to our provincial elementary and secondary school curricula, as well as a whole series of how-to videos showcasing local celebrities.

The website also displayed our collective progress in achieving our Landmark goals, displayed on a colorful graphic that was updated monthly. In addition, Landmark reports submitted through the website or app gained points for participants, which were displayed on a live Leaderboard: one for families and a separate one for community groups, such as individual school classes. This spurred excitement and a bit of friendly competition. So that no-one was left behind, the Leaderboards

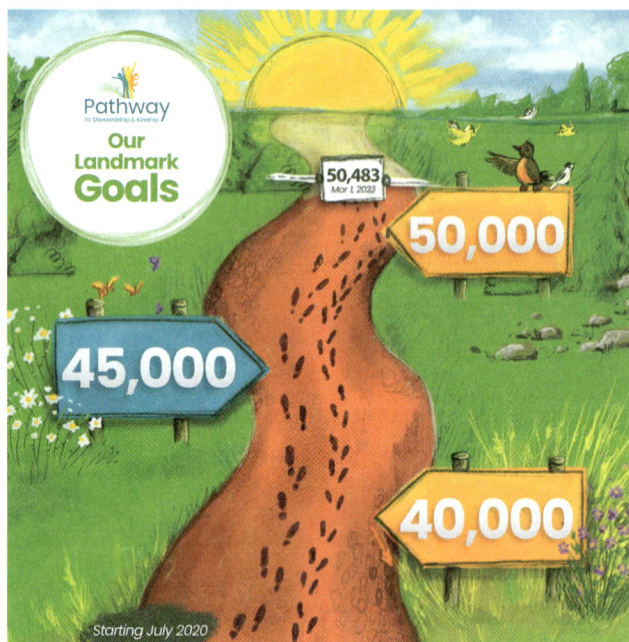

were reset to zero at the beginning of each new season (winter, spring, summer, fall), giving everyone a chance to see their progress in each season of the year.

- **Social Media:** To further distribute ideas for activities received in our Landmark reports, and build community momentum, we participated in several social media platforms: Facebook, X (formerly know as Twitter), and Instagram. Regular posts and interactions with followers created many fascinating discussions and engaged many followers who shared personal perspectives from both parents and educators. Social media was also an excellent tool for promoting special events hosted by the Pathway Project.

- **Seasonal Newsletters:** Newsletters have also become quite simple to produce and distribute through a variety of online applications. We decided to distribute newsletters four times a year, at the beginning of each new season. That way, people could stay in the loop, without being bogged down with too much material in their email boxes. We distributed newsletters to an extensive mailing list that included all participating educators, families, Roundtable members, partners, and a variety of local contacts, as well as interested people from farther afield. Anyone could ask to be on our mailing list through a simple request form on the Pathway website. We used our newsletters to showcase recent Pathway participation, new resources, upcoming events of interest, as well as Landmark-linked seasonal ideas for outdoor activities.

> I was able to go outside in winter, in conditions that I otherwise would have avoided without the prompts from the Pathway Project. Students so enjoyed the experiences.
>
> —(teacher)

- **Family Activities for School Newsletters:** Since our goal was to involve young people in Landmark-linked activities at home as well as at school and in the broader community, we tried a variety of ways of involving families. For example, we drew on each school's communication networks, one of which is sending e-newsletters to the families of children in their school. Each month, we created a poster of a fun family activity (linked to Landmarks in some way), and sent a PDF directly to participating school principals so they could easily plug it into the newsletters they were already preparing. We found they were grateful for the up-beat ideas for wholesome family fun, and were happy to distribute these posters for us. This helped us keep in touch with large numbers of families with minimal additional work for us.

● **Videos**: A unique component of the Peterborough region Pathway Project was the creation of a series of professional videos to help parents and educators develop skills and confidence in facilitating outdoor learning with the children in their care. Each of the 30 Landmarks has its own short how-to video highlighting a skilled local mentor who explains the importance of each Landmark for its target age group, as well as providing ideas and challenges for making each Landmark a part of every child's growth and development. These videos are an incomparable permanent resource on the Pathway website.

A number of our professional development workshops were also filmed and posted on the website for future reference. Each of the recorded workshops links to at least one Landmark and showcases the skills of a local mentor working with a variety of age groups.

● **Outdoor Consultants**: Throughout the development and expansion of the Pathway Project, we worked to develop personal supportive relationships with participants wherever possible. We were able to link an outdoor education expert with each participating educator who was in regular

> The Landmarks suggested have brought meaningful nature connections and discussions with my students. Following the Landmarks help me shape my units and lesson plans as well. The activities suggested through the Pathway social media and website have given me lots of ideas to try out with the class.
> —(teacher)

communication to invite discussion and questions and to offer useful resources. Wherever possible, our Outdoor Consultants visited teachers in person and shared a host of ideas for seasonally linked activities for their classes. This was especially helpful in building confidence amongst educators with less experience in outdoor learning.

> Nancy's emails with links to recommended resources were extremely helpful! I used many of these and they were all very successful with students.
> —(teacher)

> Being a consultant has enabled me to use the knowledge and experience from those 30 years of outdoor education to help others get their kids outside and caring about the natural world. It's been a wonderful extension of my working career, being able to pass along the knowledge and the joy!
> —Kim Dobson
> (outdoor activity consultant)

The very heart of the Pathway Project involves weaving together a love and appreciation for the natural world, along with developing a strong network of community care and support. These opportunities to develop supportive, meaningful personal friendships provide the foundation for personal, community, and environmental health.

- **Skill-Building Workshops**: Keeping in mind that teachers and parents are very busy people, it's still important to offer opportunities for in-person professional development wherever possible. To reach teachers, it's ideal to link to school board-led professional activity days. However, that requires a great deal of lead time and strong ties to upper management in school boards. It may be possible to access release time for teachers during the school day to attend training workshops deemed worthwhile by the school board.

Otherwise, in-person workshops can be offered on weekends for parents and teachers, or short, live online workshops can also be useful for adult skill-building. This sample of workshop topics offered by the Pathway may spark ideas for your own community:

- Wildlife, Wonder, and Weeds for primary educators
- Winter Activities for Early Learning educators
- The Hidden Life of Ponds for parents and teachers
- Nature Songs and Rhymes for parents of tiny tots
- Family Winter Night Hike
- Family Spring Hikes in local parks
- Water Quality Workshop for intermediate educators
- Nature-Based Art for parents and teachers
- Citizen Science for Intermediate educators

> I have learned so much this year through the workshops you provided us and the experts you connected us with. I especially enjoyed the ones we did virtually with Nature Nancy and Jacob Rodenburg. I feel inspired to do even more next year!
>
> —(teacher)

> I approach learning outdoors with a better sense of how to best support my students in the outdoor learning environment. I have less fear of encountering the unknown and being unable to answer questions on the spot—it's all about fostering the joy for nature.
>
> —(teacher)

- **Summer Family Adventure Challenge**: During the summer months, we launched a fun program geared for families. Each week, we sent out a challenge for a family to work on together, with ideas on how to tweak the activity for older or younger children. The challenges included a range of outdoor activities, such as finding faces in nature or exploring the night sky together. Families reported their activities on the

Pathway website or mobile app. Any family who completed and reported any five of the eight summer challenges was entered into a grand prize draw of up to $500 value in materials or services from local businesses. This generated some wonderful family activities, reported to us with heartwarming photos.

> This is a great way to spend the summer! We had so much fun looking for faces in nature.
> —Ashley Page (parent)

Problem-Solve Together

While keeping a positive outlook and sharing success stories is an important part of maximizing participation and maintaining momentum, challenges and conflicts inevitably crop up, and should be acknowledged and addressed quickly.

Sometimes, challenges are of a process-related nature, such as trying to increase access or participation in a particular sector. In this case, brainstorming potential solutions with a group of experienced people (such as a Leadership Roundtable or steering committee) can provide very helpful solutions. Other common challenges are conflicting opinions on basic principles or processes. In this case, listening carefully to all opinions is always important, and hopefully reasonable compromises can be found that are mutually agreeable. Sometimes, the disagreeing parties can respectfully agree to disagree if the conflicting issue doesn't interrupt the progress of the collective. Different per-

spectives can be a point of strength in a collective, as long as everyone can keep their eye on the shared goal of the program.

If individual personality conflicts make it difficult for certain people to work together, they can individually contribute their talents and experiences in separate working groups. An overarching guideline to bring into larger community projects is that there is strength in diversity, and a mutually respectful attitude must be brought into all interactions. If this is coupled with a sincere desire to understand the perspectives of others, a fruitful working environment can be maintained. Always avoid letting conflicts degenerate into gossip or backbiting. Cooperation for the sake of the common good is an important skill to learn at all ages and is especially important in community-building.

Celebrate Successes

In a world with too much negative press, it's important to remember that everyone likes to feel good. That's another thing we all have in common! This can be a wonderfully energizing and motivating force, if you remember to share good news at meetings, in personal interactions, and in all of your communication vehicles. These can include:
- Social media
- Newsletters
- Website features
- Blog posts
- Committee meetings
- Awards celebrations to recognize outstanding participation

Every year, we held an awards celebration, with excellent local food and entertainment, to recognize outstanding participation in the project. We gave lovely engraved wooden plaques to individuals and groups who made notable efforts to engage children in Landmark activities. Everyone enjoys recognition, and we hope these celebrations helped provide incentive to keep up the great work.

Tips for Administration and Funding

If you are an individual family or small neighborhood collective, you can easily work on integrating Pathway Landmarks into your daily lives, without need for an organizational structure or additional funding.

But, if you are planning a larger project, or even reaching out community-wide, some advice on how to organize may be helpful.

Leadership Roundtable

In our case, we organized a Leadership Roundtable of between 10 and 20 people, representing organizations as well as committed individuals who brought unique skills or community connections to the table. We tried to make sure that our three main community sectors (health, education, and environment) were equally represented. We also tried to include frontline people working directly with the public, as well as behind-the-scenes policy and decision-makers. The Leadership Roundtable was responsible for guiding the philosophy, policies, goals, and objectives of the project. Many of those on the Leadership Roundtable had already played a role in the early planning stages of the project and brought a sense of excitement and long-term commitment to the project. Another important role of Roundtable members was to report back to their home organizations to keep them informed as the project developed and grew.

We were aware that all of these people were already very busy with their own employment commitments, so we only asked them to commit to attending a Roundtable meeting once every two months. We also established some basic terms of reference which identified the size and responsibilities of the group and some basic procedural guidelines. For example, rather than require a quorum for meetings to proceed, we decided to forego the quorum requirement but give everyone at least a couple of weeks' notice of meetings and agendas and give them the opportunity to vote remotely on policy issues. That way, we weren't held up if only a few people could attend a scheduled meeting.

During COVID, we needed to meet remotely. Once we were able to meet in person again, we offered a hybrid meeting option, but recognized that being there in person would be an asset for everyone, wherever possible.

Coordinating Committee

It was helpful to form a smaller subcommittee of the Leadership Roundtable, who met more regularly (monthly) and were available to advise staff as needed. This was a group of 3 to 5 people who committed to allocating more time to guide the project. They received a small monthly stipend to recognize their added responsibility, and became incomparable advisors to the project staff.

Staff

Staffing is required to manage and co-ordinate a large community-scale program. This could come from within the participating organizations, if they have the capacity, or from outside funding. In our case, we had a total of six years of funding support from the Ontario Trillium Foundation. As various milestones were reached, we were able to apply for continued and expanded support. We hired a Project Coordinator (a co-author of this book) and a Communications Coordinator, both working roughly four days a week. In addition, we hired several Outdoor Activity consultants on an as-needed basis, to guide and support the participating educators, as well as providing a variety of online and in-person training workshops and family events.

Funding, Finances, and Legal Direction

There must be at least one organization who takes responsibility for administering finances, providing insurance and legal guidance for any community project. This is often called the "backbone organization," and they would normally be required to have charitable status and an annual audit to be able to receive funding from foundations. In our case, Camp Kawartha (an

Credit: Rose Sanderson

environmental education and leadership organization) assumed this role, and their Executive Director and co-author of this book became a member of both the Leadership Roundtable and Coordinating Committee.

A program such as the Pathway to Stewardship and Kinship could be eligible for funding from a wide variety of sectors and foundations, especially any focused on children, education, health, or the environment. The Pathway program received not only funding support from the Ontario Trillium Foundation but also very helpful advice, guidance, and resources throughout the funding cycles.

Working Groups

Often, people in the community have specific skills or interests that really help with specific aspects of the project at specific points of time. It can be helpful to assemble working groups to assist with certain short-term activities as needed. This could include things such as website development, workshop planning, researching community resources, hiring staff, and recruiting participants.

Governance Model

On the following page is a visual representation of the structure that worked well for developing, launching, and managing our community-wide Pathway program.

I am grateful to have had the opportunity to participate as a member of the Pathway to Stewardship and Kinship Leadership Roundtable. It has provided a unique platform to network and connect with others representing a broad spectrum of organizations who are all driven by their passion for stewardship out of respect and love for nature. As a collective, we create a synergy—sharing a wealth of knowledge, experiences, insights, thoughts to formulate ideas, and strategies that go beyond meeting the set goals or objectives. I am awestruck by the impact this initiative has had in establishing opportunities for children, families, and educators to learn and most importantly strengthen their connection to nature.

—Kathy Warner
(Leadership Roundtable member)

Beyond Traditional Outdoor Education

The Pathway Project expands beyond traditional outdoor education in that it combines outdoor learning and adventures with building strong, supportive relationships with people in the community. The "people" part of the story must be given plenty of care and attention too, as we imagine and craft a new direction for our collective culture. Amongst Pathway organizers, a sense of mutual respect and empathy have helped us face and manage barriers to moving forward. In a political world where continually "criticizing the other side" seems the norm, knowing that we share a common purpose (i.e., the well-being of our kids) has provided a wonderfully positive, cooperative atmosphere that is refreshing and rejuvenating. We know that everyone's voice is equally important. If these positive relationships aren't already existing in a community, it's important to take the time to develop and maintain them.

Broad Scope Is Both a Challenge and an Asset

It is complex to manage a project that spans the life of a child from birth through teenhood, since each age group has dramatically different needs and abilities. Developing resources suitable for various ages can be more manageable by building on existing programs and resources and promoting and expanding them as much as possible. There are many gems of talented people and excellent resources in every community. Having a wide range of

PATHWAY TO STEWARDSHIP & KINSHIP

governance model

Purpose Statements

Backbone Organization: Camp Kawartha is the backbone organization. They provide legal and financial responsibility for the Collaborative while ensuring compliance with their mandate and charitable objects.

Coordinating Committee: Supervise and provide support for the Coordinator(s) and oversee the day-to-day activities of the Collaborative. (3 to 5 members)

Leadership Roundtable: Represent key community sectors, and provide guidance for philosophy, policy, goals, and objectives of the Collaborative. (10 to 20 organizational or stakeholder members)

Communications Working Group (WG): Assist with the development of effective communications strategies and tools for the Collaborative.

Evaluation WG: Develop and assess evaluation strategies to monitor community impact of the Collaborative.

Community Resource Development WG: Assist in developing a database of community resources to support landmarks in the Pathway document.

Pilot Community Development WG: Assist in pilot testing of the Pathway project in selected community hubs (to be defined further as project develops).

Reporting, Support & Relationship Graphic

Credit: Brianna Salmon

community advisors is wonderfully helpful for locating them.

In most communities, you find isolated pockets of people doing great work with specific age groups or focusing on a specific interest. The broad age span of the Pathway program can help to provide a more holistic view of child and youth development. Rather than having early learning experts segregated from junior-level educators, for example, you get a glimpse of the whole picture, sharing ideas and skills across age divisions. This can help to foster a greater sense of collective community momentum, as you take important steps together toward a broad cultural shift. And that's exciting and invigorating!

Engaging Indigenous Communities

Reconciliation is an important social and environmental issue worldwide, and welcoming local Indigenous communities to participate in program planning and delivery is important to building strong and lasting relationships. However, it's equally important to realize that Indigenous communities are deeply engaged in retrieving, reclaiming, and relearning much of their own lost cultural identity, and in healing the wounds of intergenerational trauma. It is simply impossible for Indigenous communities to respond to every request they receive for an Indigenous voice.

Opening the door to sharing knowledge and opinions depends on building mutually respectful relationships. This is especially true when educators in land-based learning want to honor the teachings of their local Indigenous communities. These relationships take a great deal of time to build and maintain if they are to be more than a token of goodwill. Patience, a genuine will to understand, and respect for the wisdom that developed over millennia are important principles to keep in mind when working together. So much has been taken from

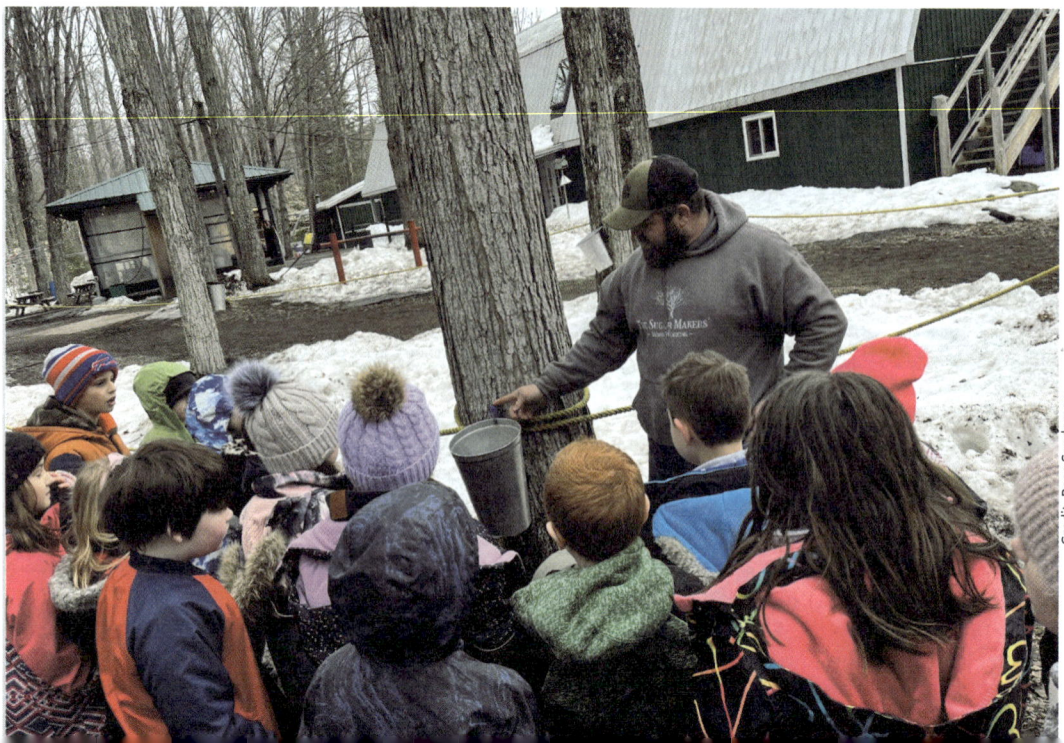

Credit: Rose Sanderson

Indigenous Peoples through colonization, and well-intentioned organizers must not continue to take from Indigenous advisors but find ways to give back as well, so we can move forward together.

We are grateful for the generosity of many Indigenous advisors throughout the development and delivery of the Pathway Project. At the same time, we realize that building a more Earth-friendly Western culture is ultimately our own responsibility, as people whose ancestors are not Indigenous to the places we live. We truly benefit from strong relationships with local Indigenous communities, but must remember that the future we create for our children depends on the choices and priorities we make today.

> We had a special visit from our friend Melody Crowe this morning. She read us a story called *When Squirrels Try*, by Lisa Galjanic. It's about a squirrel who has a very difficult time trying to get a big acorn up a tree. Finally, he asks a friend to help, and they work together to bring the acorn up. They are so happy that they were able to do it! After the story, Melody invited the children to share a time that they have been a good friend to someone. The students were so proud to share their ideas with the group.
>
> —Indrani Talapatra (teacher)

Environmental Citizenship Is Essential

We call on decision-makers in politics, education, resource management, and philanthropy to champion the creation of more opportunities for every student to engage in positive environmental initiatives. These experiences are critical for building skills and empowering youth. Many dedicated educators have already taken personal initiative and are working hard to give young students opportunities to connect with and care for the natural world. But now, we need greater systemic support and more opportunities for young people to actively protect and revitalize the communities that sustain them.

If we can inspire young people to see themselves not as passive participants in an overwhelming crisis but as capable, compassionate leaders for social and environmental change, we will have achieved something truly transformative.

The Power of Collective Action

We have shared the story of how a community can come together to guide children toward lifelong connections to caring, belonging, and stewardship. Now it's your turn to create your own story—within your family, your school, or your community. The health of our planet depends on the health of our children, and when we raise resilient, caring, and mindful young people, we all thrive. Let's work together to raise healthy children for a healthy planet!

Notes

Introduction

1. Jonathan Haidt, *The Anxious Generation: How the Great Rewiring of Childhood Is Causing an Epidemic of Mental Illness* (Penguin, 2024), 87.

2. Participaction, *The Biggest Risk Is Keeping Kids Indoors* (2015), 7. www.participaction.com/wp -content/uploads/2022/09/2015-Children-and -Youth-Report-Card.pdf.

3. Leigh M. Vanderloo, Patricia Tucker, Andrew M. Johnson, and Jeffrey D. Holmes, "Physical Activity Among Preschoolers During Indoor and Outdoor Childcare Play Periods." *Applied Physiology, Nutrition, and Metabolism* (2013) 38: 1173–75.

4. John D. Smith, David Nichols, Kyle Biggerstaff, and Nancy DiMarco, "Assessment of Physical Activity Levels of 3rd and 4th Grade Children Using Pedometers During Physical Education Class." *Journal of Research* (2009) 4: 73–79.

5. Lee Schaefer, et al., "Outdoor Time Is Associated with Physical Activity, Sedentary Time, and Cardiorespiratory Fitness in Youth." *Journal of Pediatrics* (2014) 165: 516–521.

6. Louise Chawla, "Benefits of Nature Contact for Children," *Journal of Planning Literature* (2015) 30(4): 433–452.

7. Richard Louv, *Last Child in the Woods: Saving Our Children from Nature-Deficit Disorder* (Algonquin Books, 2008), 44–47.

8. Tim Gill, "The Benefits of Children's Engagement with Nature: A Systematic Literature Review," *Children, Youth and Environments* (2014) 24(2): 10–34.

9. Rachel McCormick, "Does Access to Green Space Impact the Mental Well-Being of Children: A Systematic Review." *Journal of Pediatric Nursing* (2017) 37: 3–7.

10. Andrea Faber Taylor and Frances E. Kuo, "Children with Attention Deficits Concentrate Better After Walk in the Park," *Journal of Attention Disorders* (2009) 12: 402–409.

11. Gert-Jan Vanaken and Marina Danckaerts, "Impact of Green Space Exposure on Children's and Adolescents' Mental Health: A Systematic Review." *International Journal of Environmental Research and Public Health* (2018) 15: 2668.

12. Clara McClain and Maureen Vandermaas-Peeler, "Social Contexts of Development in Natural Outdoor Environments: Children's Motor Activities, Personal Challenges and Peer Interactions at the River and the Creek." *Journal of Adventure Education and Outdoor Learning* (2015) 16: 31–48.

13. Maria Hofman-Bergholm, "Nature-based Education for Facilitating Resilience and Well-Being Among Youth: A Nordic Perspective." *Education Sciences* (2024) 14(43).

14. Ming Kuo, Michael Barnes, Catherine Jordan, "Do Experiences with Nature Promote Learning? Converging Evidence of a Cause-and-Effect Relationship." *Frontiers in Psychology* (2019) 10: 305.

15. Louv, *Last Child in the Woods*, 86–98.

16. Christine Kiewra and Ellen Veselack, "Playing with Nature: Supporting Preschoolers' Creativity in Natural Outdoor Classrooms." *International Journal of Early Childhood Environmental Education* (2016) 4(1).

17. Catherine Ward Thompson, Penny Travlou, and Jenny Roe, *Free Range Teenagers: The Role of Wild Adventure Space and Young People's Lives*. OPENspace Research Centre: Edinburgh College of Art and Heriot Watt University, 2006.

18. Louise de Lannoy, "The Conversation: Outdoor Education Has Psychological, Cognitive and Physical Health Benefits for Children." *Outdoor Play Canada*, 2023.

19. Marcia P. Jimenez, et al., "Associations Between Nature Exposure and Health: A Review of the Evidence." *International Journal of Environmental Research and Public Health* (2021) 18(9).

20. Francisco Díaz-Martínez, et al., "Systematic Review: Neurodevelopmental Benefits of Active/Passive School Exposure to Green and/or Blue Spaces in Children and Adolescents." *International Journal of Environmental Research and Public Health* (2023), 20.

21. Jeff Mann, Tonia Gray, Son Truong, et al. "Getting Out of the Classroom and into Nature: A Systematic Review of Nature-Specific Outdoor Learning on School Children's Learning and Development." *Frontiers in Public Health* (2022) 10, 877058.

22. Gabriela Bento and Gisela Dias, "The Importance of Outdoor Play for Young Children's Health and Development." *Porto Biomedical Journal* (2017) Sep-Oct, 2(5): 157–160.

23. Ming Kuo, et al., "Do Experiences with Nature Promote Learning?" 1.

24. Jacob Petrash, *Understanding Waldorf Education: Teaching from the Inside Out.* (Gryphon House, 2002), 43.

25. Thomas Tanner, "Significant Life Experiences: A New Research Area in Environmental Education." *Journal of Environmental Education* (1980) 11(4): 20–24.

26. Nancy M. Wells and Kristi S. Lekies, "Nature and the Life Course: Pathways from Childhood Nature Experiences to Adult Environmentalism." *Children, Youth and Environments* (2006) 16(1): 1–24.

27. Louise Chawla, "Childhood Experiences Associated with Care for the Natural World: A Theoretical Framework for Empirical Results." *Children, Youth and Environments* (2007) 17(4): 144–170.

28. Ibid., 169.

29. Louise Chawla, "Pivotal Experiences in the Development of Connection and Care for Nature in Childhood and Adolescence." In *Relearn Nature* (in French translation), Cynthia Fleury and Anne-Caroline Prévot, eds. (CNRS Editions, 2017).

30. Ontario Ministry of Education, *How Does Learning Happen? Ontario's Pedagogy for the Early Years.* (King's Printer for Ontario, 2014).

31. Statistics Canada, *2012–13 Canadian Health Measures Survey*, cited in Participaction (2015).

32. John Muir, *My First Summer in the Sierra* (Houghton Mifflin, 1911).

33. A complete list of interviewees, interview questions, and reviewers can be found in the original project guidebook released in 2017. https://pathwayproject.ca/wp-content/uploads/2020/10/Pathway-to-Stewardship-Guide.pdf.

34. Forest School Canada, *Forest and Nature School in Canada: A Head, Heart, Hands Approach to Outdoor Learning* (2014). http://childnature.ca/wp-content/uploads/2016/05/FSC-Guide_web.pdf.

35. Nicole Bell, "Anishinaabe Bimaadiziwin: Living Spiritually with Respect, Relationship, Reciprocity and Responsibility." *Contemporary Studies in Environmental and Indigenous Pedagogies: A Curricula of Stories and Place* (2013) 5: 89–107.

36. Rachel Howell and Simon John Allen, "Significant Life Experiences, Motivations and Values of Climate Change Educators." *Environmental Education Research* (2019) 25(6): 813–831.

Key Stewardship Principles

37. United Nations Human Rights, Office of the High Commissioner, Convention on the Rights of the Child (1990) Article 29.1(e). www.ohchr.org/EN/ProfessionalInterest/Pages/CRC.aspx.

38. Rachel Carson, *The Sense of Wonder* (Harper-

Collins, 1998), First published in 1965 by Harper and Row.

39. Louv, *Last Child in the Woods*, 172.
40. David Sobel, "Beyond Ecophobia," *Yes Magazine* (1998). www.yesmagazine.org/issues/education-for-life/803.
41. David. A. Gruenwald, "The Best of Both Worlds: A Critical Pedagogy of Place," *Educational Researcher* (2003) 32(4): 3–12.
42. David R. Newhouse, "Ganigonhi:oh: The Good Mind Meets the Academy." *Canadian Journal of Native Education*, 31(1) (2008): 184–197.
43. Clinton L. Beckford and Russell Nahdee, "Teaching for Ecological Sustainability: Incorporating Indigenous Philosophies and Practices," *What Works? Research into Practice* (Ontario Ministry of Education, 2011).
44. W. Andrew Kenney and Danijela Puric-Mladenovic (2014), "Community Engagement in Urban Forest Stewardship: Neighbourwoods© Approach." Presented at the 11th Canadian Urban Forest Conference. http://neighbourwoods.org.
45. Ruth A. Wilson. *Nature and Young Children: Encouraging Creative Play and Learning in Natural Environments* (Routledge, 2008).
46. Chawla, "Childhood Experiences Associated with Care for the Natural World," 158.
47. Louv, *Last Child in the Woods*, 166.
48. Participaction, *The Biggest Risk Is Keeping Kids Indoors*, 7.
49. Robert D. Bixler and Myron F. Floyd, "Nature Is Scary, Disgusting, and Uncomfortable," *Environment and Behavior* (1997) 29 (4): 443–467.
50. Kristi S. Lekies, Greg Yost, and John B. Rode, "Urban Youth's Experiences of Nature: Implications for Outdoor Adventure Recreation," *Journal of Outdoor Recreation and Tourism* (2015) 9: 1–10.
51. Haidt, *The Anxious Generation*, 270.
52. Ibid., 255.
53. Antonella Rissotto and Francesco Tonucci, "Freedom of Movement and Environmental Knowledge in Elementary School Children," *Journal of Environmental Psychology* (2002) 22: 65–77.
54. Maria G. Pacilli, Ilaria Giovannelli, Miretta Prezza, and Maria Augimeri, "Children and the Public Realm: Antecedents and Consequences of Independent Mobility in a Group of 11–13-Year-Old Italian Children," *Children's Geographies* (2013) 11: 377–393.
55. Roger L. Mackett, Belinda Brown, Yi Gong, Kay Kitazawa, and James Paskins, "Children's Independent Movement in the Local Environment," *Built Environment* (2007) 33: 454–468.
56. Participaction, *The Biggest Risk Is Keeping Kids Indoors*, 8, 23–25.
57. Louv, *Last Child in the Woods*, 64–67.
58. Louise Chawla, personal email communication, March 2017.
59. Pennsylvania Land Trust, N.D., *Nature Play: Nurturing Children and Strengthening Conservation Through Connections to the Land.* naturalstart.org/sites/default/files/natureplay booklet_palta_final_print_version_pdf.
60. Susan Staniforth, *Leap into Action: Simple Steps to Environmental Action*, BC Conservation Foundation, 2003. https://hctfeducation.ca/file/leap-into-action.pdf.
61. Stan Kozak and Susan Elliot, *Connecting the Dots: Key Strategies That Transform Learning for Environmental Education, Citizenship and Sustainability.* Learning for a Sustainable Future (Maracle, 2014). https://lsf-lst.ca/media/LSF_Connecting_the_Dots_February2014.pdf.
62. Harold R. Hungerford and Trudi L. Volk, "Changing Learner Behavior Through Environmental Education," *Journal of Environmental Education* (1990) 21(3): 8–21.

Bibliography

Beckford, Clinton L. and Russell Nahdee. "Teaching for Ecological Sustainability: Incorporating Indigenous Philosophies and Practices." *What Works? Research into Practice*. Ontario Ministry of Education, 2011.

Bell, Nicole. "Anishinaabe Bimaadiziwin: Living Spiritually with Respect, Relationship, Reciprocity and Responsibility." *Contemporary Studies in Environmental and Indigenous Pedagogies: A Curricula of Stories and Place* (2013) 5: 89–107.

Bento, Gabriela and Gisela Dias. "The Importance of Outdoor Play for Young Children's Health and Development." *Porto Biomedical Journal* (2017) Sep–Oct, 2(5): 157–160.

Bixler, Robert D. and Myron F. Floyd. "Nature Is Scary, Disgusting, and Uncomfortable." *Environment and Behavior* (1997) 29 (4): 443–467.

Carson, Rachel. *The Sense of Wonder*. HarperCollins, 1998. First published 1965 by Harper and Row.

Chawla, Louise. "Benefits of Nature Contact for Children." *Journal of Planning Literature* (2015) 30(4): 433–452.

Chawla, Louise. "Childhood Experiences Associated with Care for the Natural World: A Theoretical Framework for Empirical Results." *Children, Youth and Environments* (2007) 17(4): 144–170.

Chawla, Louise. "Pivotal Experiences in the Development of Connection and Care for Nature in Childhood and Adolescence." In *Relearn Nature* (in French translation), Cynthia Fleury and Anne-Caroline Prévot, eds. CNRS Editions, 2017.

De Lanoy, Louise. "The Conversation: Outdoor Education Has Psychological, Cognitive and Physical Health Benefits for Children." *Outdoor Play Canada*, 2023. www.outdoorplaycanada.ca /2023/07/24/the-conversation-outdoor-education -has-psychological-cognitive-and-physical-health -benefits-for-children.

Díaz-Martínez, Francisco, Miguel F. Sánchez-Sauco, Laura T. Cabrera-Rivera, et al. "Systematic Review: Neurodevelopmental Benefits of Active/ Passive School Exposure to Green and/or Blue Spaces in Children and Adolescents." *International Journal of Environmental Research and Public Health* (2023) 20.

Dueck, Cathy and Jacob Rodenburg. *Pathway to Stewardship and Kinship: Raising Healthy Children for a Healthy Planet*. Pathway Project, 2017. https:// pathwayproject.ca/wp-content/uploads/2020/10 /Pathway-to-Stewardship-Guide.pdf.

Faber Taylor, Andrea and Frances E. Kuo. "Children with Attention Deficits Concentrate Better after Walk in the Park." *Journal of Attention Disorders* (2009) 12: 402–409.

Forest School Canada. *Forest and Nature School in Canada: A Head, Heart, Hands Approach to Outdoor Learning* (2014). https://childnature.ca/wp -content/uploads/2017/10/FSC-Guide-1.pdf.

Gill, Tim. "The Benefits of Children's Engagement with Nature: A Systematic Literature Review." *Children, Youth and Environments* (2014) 24(2): 10–34.

Gruenwald, David A. "The Best of Both Worlds: A Critical Pedagogy of Place." *Educational Researcher* (2003) 32(4): 3–12.

Haidt, Jonathan. *The Anxious Generation: How the Great Rewiring of Childhood Is Causing an Epidemic of Mental Illness*. Penguin, 2024.

Hofman-Bergholm, Maria. "Nature-based Education for Facilitating Resilience and Well-Being Among Youth: A Nordic Perspective." *Education Sciences* (2024) 14(43).

Howell, Rachel and Simon John Allen. "Significant Life Experiences, Motivations and Values of Climate Change Educators." *Environmental Education Research* (2019) 25(6): 813–831.

Hungerford, Harold R. and Trudi L. Volk, "Changing Learner Behavior Through Environmental Education." *Journal of Environmental Education* (1990) 21(3): 8–21.

Jimenez, Marcia P., Nicole V. DeVille, Elise G. Elliott, et al. "Associations Between Nature Exposure and Health: A Review of the Evidence." *International Journal of Environmental Research and Public Health* (2021) 18(9).

Kenney, W. Andrew and Danijela Puric-Mladenovic. "Community Engagement in Urban Forest Stewardship: Neighbourwoods© Approach." Presented at the 11th Canadian Urban Forest Conference, 2014. http://neighbourwoods.org.

Kiewra, Christine and Ellen Veselack. "Playing with Nature: Supporting Preschoolers' Creativity in Natural Outdoor Classrooms." *International Journal of Early Childhood Environmental Education* (2016) 4(1).

Kozak, Stan and Susan Elliot. *Connecting the Dots: Key Strategies That Transform Learning for Environmental Education, Citizenship and Sustainability.* Learning for a Sustainable Future, 2014. https://lsf-lst.ca/media/LSF_Connecting_the_Dots_February2014.pdf.

Kuo, Ming, Michael Barnes, and Catherine Jordan. "Do Experiences with Nature Promote Learning? Converging Evidence of a Cause-and-Effect Relationship." *Frontiers in Psychology* (2019) 10: 305.

Lekies, Kristi S., Greg Yost, and John Rode. "Urban Youth's Experiences of Nature: Implications for Outdoor Adventure Recreation." *Journal of Outdoor Recreation and Tourism* (2015) 9: 1–10.

Louv, Richard. *Last Child in the Woods: Saving Our Children from Nature-Deficit Disorder.* Algonquin Books, 2008.

Mackett, Roger L., Belinda Brown, Yi Gong, and Kay Kitazawa. "Children's Independent Movement in the Local Environment." *Built Environment* (2007) 33: 454–468.

Mann, Jeff, Tonia Gray, Son Truong, et al. "Getting Out of the Classroom and Into Nature: A Systematic Review of Nature-Specific Outdoor Learning on School Children's Learning and Development." *Frontiers in Public Health* (2022) 10: 877058.

McClain, Clara and Maureen Vandermaas-Peeler. "Social Contexts of Development in Natural Outdoor Environments: Children's Motor Activities, Personal Challenges and Peer Interactions at the River and the Creek." *Journal of Adventure Education and Outdoor Learning* (2015) 16: 31–48. www.scopus.com/inward/record.url?eid=2-s2.0-84954364318&partnerID=tZOtx3y1.

McCormick, Rachel. "Does Access to Green Space Impact the Mental Well-Being of Children: A Systematic Review." *Journal of Pediatric Nursing* (2017) 37: 3–7.

Muir, John. *My First Summer in the Sierra.* Houghton Mifflin, 1911.

Newhouse, David R. "Ganigonhi:oh: The Good Mind Meets the Academy." *Canadian Journal of Native Education*, 31(1) (2008): 184–197.

Ontario Ministry of Education. *How Does Learning Happen? Ontario's Pedagogy for the Early Years.* King's Printer for Ontario, 2014.

Pacilli, Maria Giuseppina, Ilaria Giovannelli, Miretta Prezza, and Maria Lucia Augimeri. "Children and the Public Realm: Antecedents and Consequences of Independent Mobility in a Group of 11–13-Year-Old Italian Children." *Children's Geographies* (2013) 11: 377–393.

Participaction. *The Biggest Risk Is Keeping Kids Indoors.* 2015. www.participaction.com/wp-content/uploads/2022/09/2015-Children-and-Youth-Report-Card.pdf.

Pennsylvania Land Trust Association. *Nature Play: Nurturing Children and Strengthening Conservation Through Connections to the Land*, N.D. naturalstart.org/sites/default/files/natureplaybooklet_palta_final_print_version_pdf.

Petrash, Jacob. *Understanding Waldorf Education: Teaching from the Inside Out.* Gryphon House, 2002.

Rissotto, Antonella and Francesco Tonucci. "Freedom of Movement and Environmental Knowledge in Elementary School Children." *Journal of Environmental Psychology* (2002) 22: 65–77.

Schaefer, Lee, Ronald C. Plotnikoff, Sumit R. Majumdar, et al. "Outdoor Time Is Associated with Physical Activity, Sedentary Time, and Cardiorespiratory Fitness in Youth." *Journal of Pediatrics* (2014) 165: 516–521.

Smith, John D., David L. Nichols, Kyle D. Biggerstaff, and Nancy DiMarco. "Assessment of Physical Activity Levels of 3rd and 4th Grade Children Using Pedometers During Physical Education Class." *Journal of Research* (2009) 4: 73–79.

Sobel, David. "Beyond Ecophobia," *Yes Magazine* (1998). www.yesmagazine.org/issues/education-for-life/803.

Staniforth, Susan. *Leap into Action: Simple Steps to Environmental Action*. BC Conservation Foundation, 2003. https://hctfeducation.ca/file/leap-into-action.pdf.

Tanner, Thomas. "Significant Life Experiences: A New Research Area in Environmental Education." *Journal of Environmental Education* (1980) 11(4): 20–24.

United Nations Human Rights, Office of the High Commissioner. Convention on the Rights of the Child (1990) Article 29.1(e). www.ohchr.org/EN/ProfessionalInterest/Pages/CRC.aspx.

Vanaken, Gert-Jan and Marina Danckaerts. "Impact of Green Space Exposure on Children's and Adolescents' Mental Health: A Systematic Review." *International Journal of Environmental Research and Public Health* (2018) 15: 2668.

Vanderloo, Leigh M., Patricia Tucker, Andrew M. Johnson, and Jeffrey D. Holmes. "Physical Activity Among Preschoolers During Indoor and Outdoor Childcare Play Periods." *Applied Physiology, Nutrition, and Metabolism* (2013) 38: 1173–75.

Ward Thompson, Catherine, Penny Travlou, and Jenny Roe. *Free Range Teenagers: The Role of Wild Adventure Space and Young People's Lives*. OPENspace Research Centre, 2006: Edinburgh College of Art and Heriot Watt University.

Wells, Nancy M. and Kristi S. Lekies. "Nature and the Life Course: Pathways from Childhood Nature Experiences to Adult Environmentalism." *Children, Youth and Environments* (2006) 16(1): 1–24.

Wilson, Ruth A. *Nature and Young Children: Encouraging Creative Play and Learning in Natural Environments*. Routledge, 2008.

Resources

Pathway-Related Education Resources

Pathway to Stewardship and Kinship: For more information about the Pathway to Stewardship and Kinship project, visit: www.pathwayproject.ca. You'll find high quality videos that explain each of the Landmarks, resources for parents and teachers and much more.

The Wild Path Home: An award-winning documentary about the Pathway Project. To access the trailer, go to: www.thewildpathhome.ca/home. For the full documentary, visit: https://camp kawartha.ca/resource/the-wild-path-home-video/.

Camp Kawartha Resources: A wealth of lesson plans and activities on outdoor and environmental education for every grade: Resources for Educators: https://campkawartha.ca/resources-for-edu cators/. Resources for Families: https://camp kawartha.ca/resources-for-families/.

Clubs: Search for and join a local Field Naturalist Club or a Young Naturalist Club in your area.

Project Wild, Focus on Forests: An excellent source of hands-on environmental games. You need to take the course to obtain the guide. To find out more in the United States, visit: www.fishwildlife .org/projectwild. In Canada, visit: https://cwf-fcf .org/en/explore/wild-education/?src=EL.

National Environmental Education Foundation (NEEF): NEEF is the United States' leading organization in lifelong environmental learning, creating opportunities for people of all ages to experience and learn about the environment in ways that improve their lives and the health of the planet.

North American Association of Environmental Education: A great resource for all things environmental, https://naaee.org/.

Evergreen: www.evergreen.ca. A Canadian national nonprofit environmental organization with a mandate to bring nature to our cities through naturalization projects. Evergreen motivates people to create and sustain healthy, natural outdoor spaces.

EcoSchools: A certification program that helps provide your school with practical ways to become more environmentally conscious. In the United States: https://nwf.org/ecoschools. In Canada: https://ecoschools.ca/.

Journey North: www.learner.org/jnorth. Journey North tracks the journeys of a dozen migratory species each spring. Students share their own field observations with classrooms across the hemisphere. In addition, students are linked with scientists who provide their expertise directly to the classroom. Several migrations are tracked by satellite telemetry, providing live coverage of individual animals as they migrate. As the spring season sweeps across the hemisphere, students note changes in daylight, temperatures, and all living things as the food chain comes back to life.

Canadian Journal of Environmental Education: www.edu.uleth.ca/ictrd/cjee. CJEE explores a variety of environmental topics relevant to educators.

Green Teacher: www.greenteacher.com. *Green Teacher* is a magazine by and for educators to enhance environmental and global education across the curriculum at all grade levels

Vermiculture: Cathy's Composters, www.cathys composters.com. All about how to compost with worms. Has curriculum connections, activities, and worm-related equipment to set up your own vermicomposter.

Phenology: www.drewmonkman.com. A website showcasing what is going in nature at any given time of year. Drew Monkman, well-known naturalist and educator, has a wonderful website full of detailed information about what species are active during each month of the year.

Step Outside: www.resources4rethinking.ca/en /step-outside. A wonderful selection of activities and information about what is happening in nature during each season of the year.

Andy Goldsworthy: A British sculptor who makes beautiful creations out of natural materials using only his hands. You can find evocative pictures online that inspire children to make their own creations out of found natural materials by searching "Andy Goldsworthy images."

Ecological Footprint Calculator: Calculate your footprint on the Earth in a kid-friendly way. www.footprintcalculator.org/home/en.

The Big Book of Nature Activities: https://newsociety .com/book/the-big-book-of-nature-activities/ by Jacob Rodenburg and Drew Monkman, New Society Publishers, 2017. A compendium of activities, games, and natural history notes anchored in each season of the year.

The Book of Nature Connection: https://newsociety .com/book/the-book-of-nature-connection/ by Jacob Rodenburg, New Society Publishers, 2022.

Sensory awareness activities to deepen connection to nature.

Keepers of the Earth: Native American Stories and Environmental Activities for Children: by Joseph Bruchac and Michael Caduto, Fulcrum Publishing, 1997.

Braiding Sweetgrass: by Robin Wall Kimmerer, Milkweed, 2015. *Braiding Sweetgrass* blends Indigenous wisdom with scientific knowledge to explore the deep reciprocal relationship between humans and the natural world. Through personal stories and reflections, Kimmerer emphasizes the importance of gratitude, stewardship, and learning from nature.

A Walking Curriculum: Evoking Wonder and Developing a Sense of Place (K-12): by Gillian Judson, KDP, 2018. This resource is an excellent starting point for educators to develop confidence in taking classes outdoors for themed walks in the community.

David Sobel: has written many beautiful books about connecting younger children to nature and wildness, including: *Placed Based Education, Beyond Ecophobia, Nature Preschool and Forest Kindergartens, Childhood and Nature,* and *Wildplay*.

The Last Child in the Woods: Saving Our Children from Nature Deficit Disorder: by Richard Louv, Algonquin Books, 2007. A seminal book about the impact of children's disconnection from nature.

For more information/ideas/suggestions, contact Jacob Rodenburg: jacob@campkawartha.ca

Resources for Printing

To download the following printable resources visit: https://newsociety.com/book/the-wild-path-home/

Green Sheen Color Wheel
Use this wheel to see how many shades of green you can find in nature. Use clothespins to clip small samples of what you find onto the corresponding shade of green on the wheel.

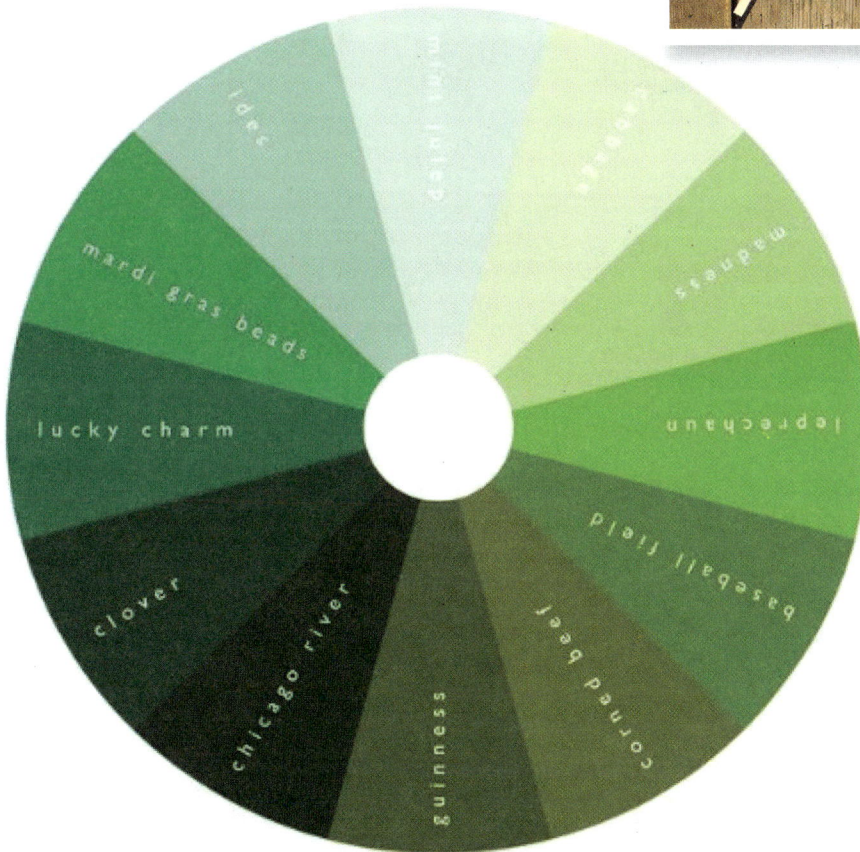

Green Sheen Colour Wheel

Seasonal Bingos
How many items can you find on the bingo card for the current season?

WINTER BINGO

How quickly can you find a line of outdoor things on this BINGO card? Can you find them all? Get out, explore and have fun!

This activity links with Landmarks 1, 2, 3, 4, 6, 9, 11, 12, 13 and 14.

Snowman	Mouse tracks	Shrub with winter leaves	Cocoon	Birds on a wire
Black and white bird	Bird nest in a forked branch	Child wearing a scarf	Animal burrow	Berries
Squirrel	Tree with white bark	FREE SPACE	Pine needles	Snowflake
Dog wearing a jacket	Three evergreen trees	Feather	Person wearing a toque	Squirrel nest
Icicles	Brightly coloured bird	Evergreen cone	Dog tracks	Birdfeeder

pathwayproject.ca

SPRING BINGO   Can you find a line of outdoor treasures on this card? Can you find them all? Explore and have fun!

Beautiful rock	Ants	Maple leaf	A flying kite	Bee on a flower
Cloud with a face	Birds feeding babies	Frog calling	Oak leaf	Bird singing
Twig with a thorn	White cedar tree	FREE SPACE	Dragonfly	Yellow flower
Bugs under a rock	People playing baseball	Caterpillar	Squirrel in a tree	Butterfly
Puddle	Yellow bird	Rain clouds	Person wearing sunglasses	Hummingbird feeder

SUMMER BINGO   Can you find a line of outdoor treasures on this card? Can you find them all? Explore and have fun!

Pathway
to Stewardship & Kinship

Caterpillar	Spider Web	Toad	Yellow flower	Birds at the top of a tree
Moss on a rock	Bird nest in a tree	Mushroom	Person with sunglasses	Red berries
Biting insect	Cedar Tree	FREE SPACE	Forked branch on the ground	Millipede
Seed on a plant	Birds in the Water	Spider	Animal making a sound	Bumble bee
Large pink rock	Hopping insect	Person wearing sandals	Ant	Leaf with jagged edges

Visit pathwayproject.ca for more ideas for outdoor fun!

AUTUMN BINGO   Can you find a line of outdoor treasures on this card? Can you find them all? Explore and have fun!

Beautiful rock	Spider	Red maple leaf	Cocoon	Birds on a wire
Black and white bird	Bird nest in a forked branch	Seed with a wing	Seed with a parachute	Berries
Squirrel hiding winter food	Tree with white bark	FREE SPACE	Pine needles	Big butterfly
Purple flower	Three evergreen trees	Feather	Acorn	Birds flying south
Pumpkin	Brightly coloured bird	Evergreen cone	Person wearing gloves	Birdfeeder

MAKE YOUR OWN BINGO   Use this blank card to challenge yourself or your family to find whatever outdoor treasures you can think of. Then, go exploring! Check off any line, or find them all. You can make a new adventure every day!

		FREE SPACE		

Index

Page numbers in *italics* indicate figures.

About the Authors

JACOB RODENBURG is an award-winning educator and executive director of Camp Kawartha, an innovative summer camp and outdoor education center. He is author of *The Book of Nature Connection* and co-author of *The Big Book of Nature Activities*. Jacob's passion is finding creative ways to connect children with nature. He lives in Peterborough, Ontario.

CATHY DUECK is a lifelong naturalist who has worked in community-based environmental education for over 35 years. She served as lead writer, researcher, and coordinator for the Kawartha Region's Pathway to Stewardship and Kinship project and has received many local, provincial and national awards. She lives in Havelock, Ontario.

ABOUT NEW SOCIETY PUBLISHERS

New Society Publishers is an activist, solutions-oriented publisher focused on publishing books to build a more just and sustainable future. Our books offer tips, tools, and insights from leading experts in a wide range of areas.

We're proud to hold to the highest environmental and social standards of any publisher in North America. When you buy New Society books, you are part of the solution!

At New Society Publishers, we care deeply about *what* we publish—but also about *how* we do business.

- This book is printed on 100% **post-consumer recycled paper**, processed chlorine-free, with low-VOC vegetable-based inks (since 2002)

- Our corporate structure is an innovative employee shareholder agreement, so we're one-third employee-owned (since 2015)

- We've created a Statement of Ethics (2021). The intent of this Statement is to act as a framework to guide our actions and facilitate feedback for continuous improvement of our work

- We're carbon-neutral (since 2006)

- We're certified as a B Corporation (since 2016)

- We're Signatories to the UN's Sustainable Development Goals (SDG) Publishers Compact (2020–2030, the Decade of Action)

To download our full catalog, sign up for our quarterly newsletter, and to learn more about New Society Publishers, please visit newsociety.com.

ENVIRONMENTAL BENEFITS STATEMENT

New Society Publishers saved the following resources by printing the pages of this book on chlorine free paper made with 100% post-consumer waste.

TREES	WATER	ENERGY	SOLID WASTE	GREENHOUSE GASES
55	4,400	23	190	23,900
FULLY GROWN	GALLONS	MILLION BTUs	POUNDS	POUNDS

Environmental impact estimates were made using the Environmental Paper Network Paper Calculator 4.0. For more information visit www.papercalculator.org

Certified (B) Corporation

new society PUBLISHERS
www.newsociety.com

FSC
www.fsc.org

MIX
Paper | Supporting responsible forestry
FSC® C016245

SDG PUBLISHERS COMPACT